FINITE ELEMENT ANALYSIS

Education and Training

This volume consists of papers presented at the First International Conference on Education and Training in Finite Element Analysis held at the Kelvin Conference Centre, West of Scotland Science Park, Glasgow, UK, 4–5 September 1991.

FINITE ELEMENT ANALYSIS

Education and Training

Edited by

JAMES T. BOYLE

Strathclyde University, Glasgow, UK

DAVID K. BROWN
BILL MAIR

Glasgow University, Glasgow, UK

PHIROZE MEHTA

Glasgow Polytechnic, Glasgow, UK

and

JIM WOOD

Paisley College, Paisley, UK

ELSEVIER APPLIED SCIENCE
LONDON and NEW YORK

ELSEVIER SCIENCE PUBLISHERS LTD
Crown House, Linton Road, Barking, Essex IG11 8JU, England

Sole Distributor in the USA and Canada
ELSEVIER SCIENCE PUBLISHING CO., INC.
655 Avenue of the Americas, New York, NY 10010, USA

WITH 16 TABLES AND 78 ILLUSTRATIONS

© 1991 ELSEVIER SCIENCE PUBLISHERS LTD

British Library Cataloguing in Publication Data

Finite element analysis: education and training.
I. Boyle, James T.
515.353

ISBN 1-85166-706-7

Library of Congress CIP data applied for

Printed in Great Britain at the University Press, Cambridge.

Introduction

Some of the early roots of numerical methods for solving technological problems go back to the first half of the 20th century. Solution of differential equations governing complex problems which could not be solved analytically were cast into difference equtions and finite difference techniques employed over a gridded region. Early work done in Glasgow by Professor Alexander Thom in the 1930s was taken up by workers such as Southwell and the relaxation techniques developed using mechanical calculators and teams of tireless researchers to produce solutions. Solutions using stress functions and the biharmonic equation are one of the many approximate techniques adopted by engineers and scientists into the 1940s.
The 1950s saw the development of a new technique with an approach from the governing integral equation and looking at solutions to discrete areas and not along lines of a grid. The term *finite element* was first used by Clough in a 1960 paper. During the 1950s and 1960s the finite element method of numerically producing approximate solutions to redundant and complex continuum problems was in the realms of the researchers and formed the substance of many PhD theses. Professors Zienkiewicz and Argyris are names which predominate at this time of rapid development. The electronic computers, which during the 1940s were emerging, started to be more readily accessible through high level languages and have more power to solve significantly sized problems. From this time can be seen the development of the first finite element packages , drawing together the specific programs developed in universities and colleges to solve specific problems. The business of developing and selling finite element packages was now clear and the commericial potential and industrial use was now going to be exploited.

In the 1960s the undergraduate teaching was traditional but was soon awakening to the need to introduce the finite element technique to students. Initially it was introduced into final year honours courses as a special topic but soon it was clear that so powerful was the technique that a full integration into the curriculum was essential. Some of the difficulties arose from the scarcity of computer resources; mainframe computing with free access was not readily available to large numbers of students.

The 1970s and 1980s have seen the provision of significant and relatively cheap computer resource at mainframe level and more importantly the availability of microcomputers, which have the power of the mainframes of the early 1960s. The educators and trainers of the finite element technique have now the difficulty of stripping out the old numerical and analytical methods (elegant as they were) and building new syllabuses and courses to streamline the introduction of finite element analysis. The computing power of the 1990s should reduce all hardware restrictions to a minimum but care must now be exercised to give the students of the 1990s a good understanding of the power of the techniques and the responsibility they have in using it.

It is from the above background that the suggestion of running a conference in the education and training in finite element analysis emerged from the Scottish Computational Engineering Group. The four HEIs (Higher Educational Institutions) in the West of Scotland - Universities of Glasgow and Strathclyde, Paisley College and Glasgow Polytechnic - combined to organise the conference along with NAFEMS (National Agency for Finite Element Methods and Standards), which is centred at the National Engineering Laboratory at East Kilbride, south of Glasgow. Support has also come from other sources - notably Scottish Enterprise, which has paid for the cost of publishing the proceedings.

The aims of the conference can be thought of as providing a forum in which finite element practitioners can meet and share experiences in the education and training of students and through publication allow a wider audience to have access to this fund of experience. It is a celebration of not knowing but willing to learn from other people things which have been tried and succeeded or failed. What do the employers expect of graduates in the way of acquired analytical, numerical and computational skills and what knowledge and understanding do educationalists expect students to have when thay graduate into the world of employment? What amount of core knowledge must be understood before students are given access to powerful computer packages? What checking techniques should be given to the students to allow for quality assurance of computer results? How much hands-on time should be given during education in the basic techniques or training in the use of specific packages?

The Conference presents a wide range of papers relating experiences of educators and trainers in this all-important technique. Minimal editing has been exercised to allow a free range of expression by the authors. The majority address the areas of structural mechanics in mechanical, civil and structural engineering as would be expected from the summary of results given by Anderson & Martin. Both overviews and particular examples with milling cutters and truck brackets are presented. Intriguing titles such as "Better guessing the stressing", "An FEM course from cradle to grave", "Computational mechanics -Chapter 20 or Appendix B?", and "Teaching finite elements to disadvantaged students" promise interesting listening and reading. Software vendors and industrialists are also represented, bridging the education training gap which is also being addressed by papers looking at integrating FEA with traditional mechanics and laboratory work. What is the core knowledge required? Is using elements without work or patch tests the way forward? Is introducion of PC-based small programs or packages such as CALFEM the answer or should we go straight onto large mainframe based packages such as PAFEC or ABAQUS? All these issues are dealt with in the set of proceedings. They are not exhaustive but it is hoped that by sharing their experiences others will benefit and try out new ideas.

Where does the future lie for educating the finite element practicioners of the 21st century? NAFEMS stands to ensure quality of provision of both software vendors and users. The pointers look towards to provision of more comprehensive computer packages which can model, solve, optimise and even finally instruct machine tools to produce final widgets! Anderson & Martin's "FE2000" gives a checklist of future requirements. However we must ensure that there are the users who understand the processes, can safely drive the software tools and also have engineering judgement to assess the validity of the results. Educationalists and trainers alike must constantly revise course content to ensure a balance between core knowledge and user understanding of the techniques being employed.

It is intended that this conference is the first of a series addressing these above topics and providing a regular forum for the sharing of experiences from both industry and academia.

The editors wish to gratefully acknowledge the support of

Scottish Enterprise

Institution of Mechanical Engineers

Technology Training Partnership

The Design Council

Contents

AN ACADEMIC VIEW OF FINITE ELEMENT EDUCATION AS PREPARATION FOR FUTURE PRACTICE

D. R. J. OWEN
Department of Civil Engineering
University College Swansea, U.K. SA2 8PP

INTRODUCTION

The finite element method is now firmly established as a design tool in many branches of engineering. No longer is it regarded as the sole prerogative of researchers but is used for the day to day analysis of a variety of components and structures. Integration of finite element based packages with CAD systems provides an efficient design route and most problems can be solved on relatively low cost hardware. The power of present day workstations permits the solution of even large scale problems and the man/machine interaction provided by such equipment makes its use attractive. Even the latest 32 bit micro or desktop computers are sufficiently powerful to undertake substantial finite element calculations and their low cost puts computational design capabilities within the reach of even the smallest design office.

The almost universal popularity of the finite element in analysis and design has lead to the widespread use of commercial packages throughout the engineering industry. Whilst every effort is made by software developers to ensure that such a "black box" approach can be safely employed by engineers with no special finite element experience, it is essential that users should have a sound basic knowledge of the principles of the technique and its application to engineering problems. In particular, users must have an appreciation for discretisation principles, element performance, modelling procedures and material constitutive behaviour since the validity of any finite element solution will depend on a rational consideration of these factors.

There is no single "correct" way of imparting such knowledge to users or students, since the most efficient approach clearly depends on issues such as background, previous experience and field of application. Therefore the contents of this paper will be based on the author's personal experience in teaching finite element methods to undergraduate and postgraduate students and providing training courses to engineers in a professional environment. It is proposed to examine separately the problems arising from the teaching of finite elements to undergraduates, postgraduate students and professional engineers although many issues are common. Some discussion will also be provided on future trends in finite element development and educational requirements.

UNDERGRADUATE TEACHING

A basic problem facing academic establishments is that the traditional undergraduate degree cirriculum is already overcrowded, making the introduction of new topics, such as the finite element method, a very difficult task. It is tempting to consider the generation of curriculum space by the removal of some of the more classical and analytical elements of the course - based on the rationale that numerical methods are now replacing such traditional approaches. However it can be equally argued, with the foundations of the finite element approach being firmly based on energy and mathematical principles, that fundamental studies take on an increasing importance. In the present academic financial climate it is difficult to justify the teaching of finite elements on a subject option basis. Indeed, in view of its relevance to modern engineering practice it should form a compulsory part of the undergraduate curriculum. The clear way forward would appear to be offered by the universal introduction in UK Universities of a four year undergraduate degree course to both bring the UK more in line with European institutions and allow adequate time for the teaching of additional subjects.

At University College Swansea, the Department of Civil Engineering is the only Engineering Department to offer a comprehensive finite element course at undergraduate level. This comprises 25 one hour lectures with one hour/week examples or practical sessions. These formal tuition periods are complemented by individual student projects.

The teaching of finite elements to Civil Engineering students is possibly simpler than to students of other engineering disciplines in view of their previous exposure to matrix methods of structural analysis. This computational technique for the analysis of frameworks contains most of the steps to be found in a finite element solution and it is only the discretisation process that has to be additionally introduced.

The structure of the finite element course taught at Swansea is summarised in Table 1 and, to the author's view has the following important features:-

Computational Analysis and Relation to Design Procedures

It is essential to place the finite element method in its proper perspective with regard to its role in engineering design. This is introduced by describing a computational model in the form illustrated in Fig. 1. Consider an engineering problem which might, for example, involve the selection of an economic design for a particular structure or component. A numerical solution sequence must include the following considerations:

Appropriate Modelling of the Problem. Today's powerful solution capabilites permit great flexibility with regard to the modelling of structural behaviour. For example, nonlinear and/or time dependent effects may be included in the material description. It is important that the designer chooses the most appropriate constitutive behaviour and, equally important, that he does not over-elaborate on the choice of model.

The designer must make a further choice on appropriate modelling of the structural action. For example, is a beam or plate bending representation sufficient? Or does the complexity of the load carrying action warrant a full three dimensional analysis?

The proper modelling of an engineering problem is a most important and difficult task and often requires considerable ingenuity on the part of the analyst. It is frequently claimed that computers kill ingenuity - in fact the converse can and indeed should hold.

Applicability of Numerical Techniques. Any innovative numerical process should be developed to such an extent that it can be employed with confidence by engineers in analysis and design. This requires the comparison of numerical and experimental results and application of the process to practical problems.

Associate Software Development. Numerical algorithms must be implemented in good quality, reliable, user-friendly and well documented software which is readily applicable on appropriate computers for use by engineers and research workers. This software must be verified and the user educated in its correct use. Of particular importance in this respect are quality assurance issues.

Relation to Design Procedures. A numerical analysis is more often than not an integral part of engineering design. Therefore it is important that the results of a numerical computation can be interpreted in an unambiguous manner; especially in view of the approximate nature of computational models.

One-dimensional Models

Many of the features of finite element analysis can be taught through one-dimensional models. This has the advantage of eliminating much non-essential detail as well as reducing the amount of computation required for demonstration examples. In particular the following components of the method can be described:

  *The discretisation process and shape functions

*Formulation of element "stiffness" matrices and "load"
 vectors
*Global assembly processes
*Equation solution techniques

Examples in this context include heat conduction in one-dimension and stress analysis in axial bar systems.

Quasi-Harmonic Problems

It is particularly instructive to teach the finite element method for two dimensional situations through application to the quasi-harmonic equation for the following reasons:-

*There is only one unknown at each nodal point. This eliminates the complexities introduced by multi-variable problems which add little to the understanding of the finite element process.

*Development of the appropriate discretised equations and associated numerical solution procedures allow the solution of a wide range of practical engineering problems; some of which are listed in Table 2.

Of course, extension to stress analysis problems which involve more than one degree of freedom per node is necessary; for both Civil and Mechanical Engineers at least.

Execution of Hand Examples

The complete solution for some small numerical examples by hand is essential for a complete understanding of the basic principles. A typical example is included in Appendix I. Such problems give the student experience and confidence in the evaluation of stiffness matrices and force vectors and assembly of element contributions to form the global equations. By appropriate choice of material parameters, round-off errors can be demonstrated if the equation system is solved using the same truncation principles employed by computers.

Commercial Codes

The solution of some relatively simple problems using commercial software is desirable. This introduces students to the solution of problems beyond simple hand examples and should serve as an introduction to professional practice. However the codes should be extremely user friendly and the extent of this activity must of necessity be restricted in view of curriculum pressures. At Swansea the MICROFIELD system developed by Rockfield Software Ltd. is used for this purpose.

POSTGRADUATE TRAINING

The level of postgraduate instruction clearly depends upon the emphasis placed upon finite element techniques in the research strengths of the institution concerned. At Swansea, considerable research effort is devoted to the development of finite element processes which is consequently reflected in the M.Sc. course offered by the Department of Civil Engineering. Finite element problems can generally be divided into three categories.

Equilibrium problems in which no variation with time takes place (e.g. steady-state heat conduction, linear and nonlinear structural analysis).

Eigenvalue problems. These are extension of equilibrium problems in which, due to the system properties, solutions exist only for critical values of certain parameters (e.g. free vibration analysis and buckling problems).

Propagation problems in which some time-dependent phenomenon takes place (e.g. heat conduction under transient conditions, slow transient viscoplastic deformation and dynamic transient effects such as seismic or impact loading).

The extent to which the above topics are covered in any M.Sc. course is dictated by time, but in view of the now almost universal acceptance of commercial codes offering nonlinear solution capabilities, it is essential that a thorough introduction to nonlinear solution processes be covered.

The step from linear to nonlinear solution algorithms requires a major advance in both underlying theory (hardening plasticity, unified viscoplasticity, finite strain formulations, contact-friction, etc.) and numerical implementation processes (consistent tangent formulations, multi-step time incrementation techniques, automatic load incrementation, etc.). Therefore this topic usually comprises the major part of any advanced finite element course.

As well as formal tuition in the theory and solution processes it is essential that students be given practical experience in the use of such techniques. This is even more important than for linear problems, since the solution of nonlinear problems inevitably requires considerably more insight to the physical behaviour. Students should be introduced to well structured and well documented codes and the practical component of the course should lead to the final position where they are capable of assembling a complete package (employing standard routines or a shell) for a particular nonlinear application.

Modern trends in finite element analysis are leading towards the use of adaptive refinement principles and commercial codes are beginning to feature such facilities. Error estimation, which is the essential ingredient of adaptive refinement, is of crucial importance in appraising solution integrity and is a fundamental topic which should be included in any advanced finite element course. The development of adaptive meshes, based on an estimate of the solution error, is however a more esoteric topic and cannot be readily covered in a general course.

CONTINUING EDUCATION AND PROFESSIONAL TRAINING

The rapid progress that continues to be made in the development of finite element techniques and which are subsequently implemented in commercial codes requires the periodic training of practicing engineers. This can be accomplished in several ways:-

Postgraduate industrial training

The IGDS (Integrated Graduate Development Scheme) has recently been introduced whereby practicing engineers can study towards an M.Sc. degree from within their company. The basic idea is that courses are offered as one week modules which can be taken at intervals until sufficient credits have been accumulated. This course work is complemented by a project of direct relevance to the participant's own organisation. In this way advanced training can be obtained with minimum disruption to the company's activities.

Short training courses

A survey of the trade literature will indicate that a wide range of finite element courses are currently being offered, which range from general or background information courses to specialist courses on focussed topics. Some courses are also offered on an "in-house" basis to minimise a company's operational disruption. The prime intention of courses of this type is usually to educate finite element users in the safe operation of commercial codes. In this respect both code vendors and users have distinct responsibilities:-

> Vendors must make every effort to validate their codes through extensive benchmarking and the inclusion of features such as automatic data generation and display, error estimation and adaptive refinement, automatic load incrementation for nonlinear problems, etc.

> Users must not adopt a "black box" approach to solution. It is essential that they have a knowledge of the underlying theory, the element methodology and the numerical solution techniques employed, since it is important that they know the limitations of the solution process (the vendor will usually have emphasised the positive aspects). It is also important that users undertake their own benchmark tests, both to ascertain that the code has been correctly installed and to gain confidence in its use.

In the context of benchmarking and code validation the role of NAFEMS should be highlighted and its activities will assume even greater prominence as the issue of quality assurance of finite element codes and their usage becomes further established.

Distance Learning Procedures

There is increasing interest in the development of distance

learning utilities for finite element analysis. To date there have been some contributions in this direction through the SOCRATES and ATHENA projects, based at Cornell University and M.I.T. respectively, but these have not been specifically aimed at the finite element method. A project is currently in progress, sponsored through the COMETT initiative of the EC, whose specific objective is to develop PC or workstation based software for finite element education. The first aim of this AMADIS project is to provide distance learning packages for a basic finite element course which will be later complemented by more advanced modules.

FUTURE TRENDS

Finally, it may be appropriate to examine the effects on finite element education of future trends in the development of computational techniques. Some of the current aims are towards automating finite element analysis and design procedures, thereby removing many of the decision making processes from the user. It can be argued that if such objectives are fully met, then the need for a rigorous finite element education could be diminished. The main developments in this direction are perceived to be:-

Artificial Intelligence and Expert Systems
Artificial intelligence is considered to be the cornerstone of the next generation computers. Steady progress in this field has been made for the last 20 years or so and has led to the present day position where machines are available capable of heuristic (rule of thumb) reasoning based on a limited data bank of knowledge. (At the very simplest level we could cite the example of electronic chess games). Artificial intelligence functions essentially by applying a sequence of IF....THEN logic operations to a number of stored items of information; typically 10,000, using a special processing computer language such as LISP. Before artificial intelligence can be used to the degree of sophistication envisaged for future engineering applications, the development of fifth generation machines, which have the following abilities, is necessary:-

Object detection. Next generation machines should be able to detect objects from given pictorial images. Some advances are already being made on this front following at least two independent approaches.

Reasoning ability. For the development of so-called "expert systems" the computer will require the ability to reason.

A present day cliche in the engineering world is that "computers are good drudges but no judges". The aim of expert systems is to make this statement obsolete. An expert system can be defined as an "user friendly" data base which stores the knowledge and experience of a host of experts. In this way it is envisaged that engineering design tasks currently requiring the services of a professional engineer could be performed by anyone with access to such a system. It is not expected that comprehensive expert systems, incorporating finite element analysis modules, will be generally available much before the end of the century by which time systems with the following capacities will have been developed:

*A living system in continual transition in that the knowledge database is being progressively upgraded.

*The system should be able to learn from its own previous errors and experience.

*The user can interact with the system - the system may even learn and account for the user's idiosyncracies.

*The system will have no human hangups. In particular pride or ego will be absent and the system will also not suffer from emotional problems.

The above specification summarises the dream - the reality may possibly turn out to be something quite different:

*The knowledge input may be incomplete leading to incorrect decisions being made in inexperienced hands.

*The output may be sufficiently unreliable to require expert interpretation.

*There may always be a need for engineers to have a detailed knowledge of the software performance.

*For finite element programming the use of unstructured languages, such as the FORTRAN language used by most engineers, may still be unavoidable.

Pre and Post Processing

The easy and automatic preparation of input data for complex finite element analysis still remains a goal. Such systems would include a direct link to CAD data for geometry definition and automatic mesh generation, including adaptive refinement procedures for producing near optimal meshes.

Additional developments in automatic load incrementation and time stepping procedures will remove further decision making from the user and the realisation of some of the promises from the field of shape optimisation will provide direct benefits for economic design.

In summary, it can be seen that several of the future developments in the field of finite elements will lead to a more automatic usage with fewer user decisions being necessary. However, it is difficult to conclude at the present time that such advancements will diminish the need for a fundamental education in the principles and implementation of the finite element method.

Table 1   Undergraduate finite element course curriculum

Topic	Lectures
Introduction - Relation of Finite Elements to design procedures	2
Review of matrix methods for structural analysis and equation solution techniques	2
Energy methods, Rayleigh-Ritz techniques and Weighted Residual Approaches	3
Discretisation procedures and shape functions	2
Finite elements in one-dimensions *Axially loaded bar systems *Heat flow in a slender road *Timoshenko beam analysis	5
Finite elements in two-dimensions Quasi-Harmonic problems	5
Advanced two-dimensional concepts *Isoparametric elements *C(1) and related elements	2
Solid mechanics applications	4

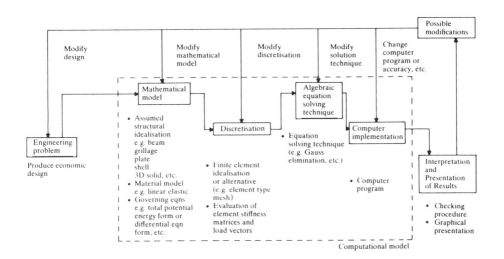

Fig 1   A finite element computational model

Table 2   Physical situations governed by the quasi-harmonic equation

Physical problem	Unknown, ϕ	K_x, K_y	Q
Heat conduction	Temperature	Conductivity	Internal heat generated
Gas Diffusion	Concentration	Diffusivity	
Seepage	Pressure head	Permeability	
Compressible flow	Velocity potential	Density	
Magnetostatics	Magnetic potential	Reluctivity	Current density
Torsion	Stress function	(Shear modulus)$^{-1}$	Twist
Torsion	Warping function	Shear modulus	
Reynolds film lubrication	Pressure	(Film thickness)^3/ viscosity	Lubricant supply

APPENDIX I

ILLUSTRATIVE EXAMPLE OF THE SOLUTION OF A TWO-DIMENSIONAL QUASI-HARMONIC PROBLEM BY FINITE ELEMENTS

<u>Problem Definition</u>

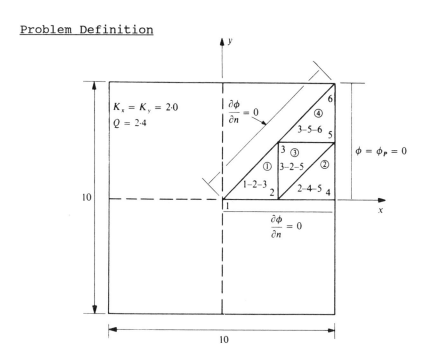

<u>Element Stiffness Matrices and Force Vectors</u>

$$
K^{(1)} = K^{(2)} = K^{(4)} = \begin{bmatrix} 1 & -1 & 0 \\ -1 & 2 & -1 \\ 0 & -1 & 1 \end{bmatrix}, \qquad K^{(3)} = \begin{bmatrix} 2 & -1 & -1 \\ -1 & 1 & 0 \\ -1 & 0 & 1 \end{bmatrix},
$$

$$
f^{(1)} = f^{(2)} = f^{(3)} = f^{(4)} = 2 \cdot 5 \begin{bmatrix} 1 \\ 1 \\ 1 \end{bmatrix}.
$$

Assembly Process

$$
\begin{bmatrix}
k_{11}^{(1)} & k_{12}^{(1)} & k_{13}^{(1)} & 0 & 0 & 0 \\
k_{21}^{(1)} & k_{22}^{(1)}+k_{11}^{(2)}+k_{22}^{(3)} & k_{23}^{(1)}+k_{12}^{(3)} & k_{12}^{(2)} & k_{13}^{(2)}+k_{23}^{(3)} & 0 \\
k_{31}^{(1)} & k_{32}^{(1)}+k_{21}^{(3)} & k_{33}^{(1)}+k_{11}^{(3)}+k_{11}^{(4)} & 0 & k_{13}^{(3)}+k_{12}^{(4)} & k_{13}^{(4)} \\
0 & k_{21}^{(2)} & 0 & k_{22}^{(2)} & k_{23}^{(2)} & 0 \\
0 & k_{31}^{(2)}+k_{32}^{(3)} & k_{31}^{(3)}+k_{21}^{(4)} & k_{32}^{(2)} & k_{33}^{(2)}+k_{33}^{(3)}+k_{22}^{(4)} & k_{23}^{(4)} \\
0 & 0 & k_{31}^{(4)} & 0 & k_{32}^{(4)} & k_{33}^{(4)}
\end{bmatrix}
$$

$$
\times
\begin{bmatrix}
\phi_1 \\ \phi_2 \\ \phi_3 \\ \phi_4 \\ \phi_5 \\ \phi_6
\end{bmatrix}
=
\begin{bmatrix}
f_1^{(1)} \\
f_2^{(1)}+f_1^{(2)}+f_2^{(3)} \\
f_3^{(1)}+f_1^{(3)}+f_1^{(4)} \\
f_2^{(2)} \\
f_3^{(2)}+f_3^{(3)}+f_2^{(4)} \\
f_3^{(4)}
\end{bmatrix}
$$

Equation Solution

Equation System

$$\phi_1-\phi_2+0\,.\,\phi_3+0\,.\,\phi_4+0\,.\,\phi_5+0\,.\,\phi_6=2{\cdot}5$$

$$-\phi_1+4\phi_2-2\phi_3-\phi_4+0\,.\,\phi_5+0\,.\,\phi_6=7{\cdot}5$$

$$0\,.\,\phi_1-2\phi_2+4\phi_3+0\,.\,\phi_4-2\phi_5+0\,.\,\phi_6=7{\cdot}5$$

$$0\,.\,\phi_1-\phi_2+0\,.\,\phi_3+2\phi_4-\phi_5+0\,.\,\phi_6=2{\cdot}5+r_4;\quad \phi_4=0$$

$$0\,.\,\phi_1+0\,.\,\phi_2-2\phi_3-\phi_4+4\phi_5-\phi_6=7{\cdot}5+r_5;\quad \phi_5=0$$

$$0\,.\,\phi_1+0\,.\,\phi_2+0\,.\,\phi_3+0\,.\,\phi_4-\phi_5+\phi_6=2{\cdot}5+r_6;\quad \phi_6=0.$$

Forward Elimination

$$\phi_1-\phi_2+0\,.\,\phi_3+0\,.\,\phi_4+0\,.\,\phi_5+0\,.\,\phi_6=2{\cdot}5$$

$$0\,.\,\phi_1+3\phi_2-2\phi_3-\phi_4+0\,.\,\phi_5+0\,.\,\phi_6=10{\cdot}0$$

$$0\,.\,\phi_1+0\,.\,\phi_2+2{\cdot}6666\phi_3-0{\cdot}66666\phi_4-2\phi_5+0\,.\,\phi_6=14{\cdot}166$$

$$0\,.\,\phi_1+0\,.\,\phi_2+0\,.\,\phi_3+1{\cdot}4999\phi_4-1{\cdot}5000\phi_5+0\,.\,\phi_6$$
$$=9{\cdot}3748+r_4;\quad \phi_4=0$$

$$0\,.\,\phi_1+0\,.\,\phi_2+0\,.\,\phi_3+0\,.\,\phi_4+2{\cdot}5000\phi_5-\phi_6=18{\cdot}124+r_5;\quad \phi_5=0.$$

$$0\,.\,\phi_1+0\,.\,\phi_2+0\,.\,\phi_3+0\,.\,\phi_4+0\,.\,\phi_5+\phi_6=2{\cdot}5+r_6;\quad \phi_6=0.$$

Backsubstitution

$$\phi_6 = 0; \quad r_6 = -2 \cdot 5,$$

$$\phi_5 = 0; \quad r_5 = -18 \cdot 124,$$

$$\phi_4 = 0; \quad r_4 = -9 \cdot 3748,$$

$$\phi_3 = \frac{14 \cdot 166}{2 \cdot 6666} = 5 \cdot 3123,$$

$$\phi_2 = \frac{10 \cdot 0 + 2(5 \cdot 3123)}{3} = 6 \cdot 8748,$$

$$\phi_1 = 2 \cdot 5 + 6 \cdot 8748 = 9 \cdot 3748.$$

Equilibrium Check

$$QA = 30 \approx 29 \cdot 9988,$$

INTEGRATION OF FINITE ELEMENT STRESS ANALYSIS WITH TRADITIONAL STRENGTH OF MATERIALS LABORATORY TEACHING

J.F.L. Chan
*Department of Mechanical, Materials, and
Manufacturing Engineering,
University of Newcastle upon Tyne,
The United Kingdom, NE1 7RU.*

ABSTRACT

This paper presents a proposal of a coherent structure for the teaching of stress analysis at the undergraduate and postgraduate level. The experimental, numerical and analytical approaches to stress analysis are described and their links to laboratory work in materials testing is indicated, together with the implications for the teaching of design.

The method adopted for the integration of the teaching in finite element techniques with the more traditional topics will be presented. This involves mini-projects with a range of specimens subjected to loads and being analyzed using both experimental and numerical techniques. Where appropriate, both the boundary element and finite difference methods will also be introduced. The projects and the examples used are also supported by separate case-study material.

The approach allows the development of concepts such as fracture, fatigue, and failure mode, effect, and criticality analysis (FMECA) to be developed and a link to be made with applications of stress analysis in design.

Throughout the modern role of finite element techniques are indicated and their utility emphasized.

INTRODUCTION

The method for teaching **strength of materials** at junior undergraduate level, and later **stress analysis** at the senior level adopted by many teachers has been the traditional "chalk and talk" approach for a good many years. This is equally true for standard experimental work which is fundamental to the understanding of the basic principles of mechanics (e.g. Holman [1]), which has been traditionally carried out almost without exception in all engineering mechanics laboratories. However, the availability of powerful workstations and indeed microcomputers, which can be easily brought into the classrooms and laboratories, has drastically changed the way in which the subject of stress analysis can be presented in classrooms and lecture theatres (Adeli & Chen [2], and Potts & Oler [3]). Formal procedures for certain fundamental analysis, such as the construction of free-body diagrams (Roth [4]), can now be presented, and can be readily incorporated into computer programmes for numerical analysis. A natural progression which follows is of course to integrate the analytical teaching with both numerical and experimental stress analysis (e.g. Dixon [5]).

Research findings on the combination of the **finite element method** (FEM) and traditional experimental techniques such as photoelasticity (Berghaus [6]), Moiré interferometry (Morton et al [7]) have been reported. Other hybrid experimental-computational approaches for "photomechanics" techniques such as holographic interferometry, speckle interferometry, and shearography (McKelvie [8]), where the data obtained by measurement are immediately available as input data in computational analysis, using for example both the finite and the boundary element methods, have also been the subjects of active research. Some modern curriculum in materials engineering has also been developed, with numerical and experimental fracture and stress analysis forming a major proportion of the undergraduate curriculum (e.g. Blicblau [9]).

The need for a coherent course structure for the teaching of stress analysis which integrates with the important subject areas of design, quality assurance, and product liability etc initiated a thorough rethinking and redesigning of both the undergraduate and postgraduate curriculum at Newcastle. The rationale, the preparation and implementation, and the limitations of such a course are presented in this paper.

ANALYTICAL, NUMERICAL, AND EXPERIMENTAL STRESS ANALYSIS

The analysis of stress at a point, either by using the Mohr's circle or the theory of elasticity can be carried out easily on relatively simple structural shapes. For both stress and design analysis on more complex structures with, say, non-uniform cross-sections which will also involve complicated loading and boundary conditions, such as bending, torsion, and thermal loading, numerical and experimental methods will have to be employed. Analytical solutions provide a standard against which the performances of both numerical and experimental methods can be judged. Throughout a course in stress analysis the modern role of numerical methods such as the finite element methods, as well as the utility of the modern experimental techniques (e.g. [6, 7, 8, 10, 11]) must be emphasized.

Strength of Materials Laboratory - Part I

The subject of finite element analysis is not included in our Part I curriculum. It is generally felt that although the finite element concept can be introduced at Part I, it would be much better to cover the subject at Part II, when the students would have a better understanding of the engineering sciences. The aims for the laboratory are therefore simply to add to the students' understanding of the subject by demonstrating the topics which they will not have come across during the lectures, and to reinforce the understanding of the course materials by providing the students with "hands on" experience in calibrating and using their experimental instruments and equipment, and to proving or confirming the principles of mechanics to which they have been introduced during the lectures.

Strength of Materials Laboratory - Part II

Students work in groups of three and attend the laboratory for three consecutive Tuesday afternoon sessions, each of which is of three hour duration. Every student on the Mechanical Engineering course will have an opportunity to work in the laboratory. The objectives of the laboratory are three fold:

- To carry out a stress analysis on a chosen component by using strain gauges, photoelasticity techniques, and the finite element or boundary element methods. Both numerical and experimental analysis must be carried out, however, any one or a combination of the methods in each of the above categories may be used.

- To attend demonstrations on experiments such as plastic bending, asymmetric bending, and other relevant tests.

- To study the failure probability of the component with regard to tests which

3

are to be carried out in order to assess material fracture toughness, impact strength and other relevant parameters.

The students are given a list of components from which each group can choose one for analysis. The components can be made of aluminum, steel, and plastics. Some typical examples of which are:

- a component with a hole in a plate
- a wedge shaped cutting tool
- a rectangular straight beam
- an eye-bar
- an eccentric cylinder
- a notched plate, two gear teeth in mesh
- a curved beam
- a flat punch on a rectangular plate
- a disc wheel, and

Session I involves:

- The selection of a component, and defining the objectives for the experiments.

- An initial analysis, e.g. Where to load the component? How much loading can be safely applied? Where to put the strain gauges? etc, will be carried out with the help from the laboratory supervisor.

- An estimation of the order of magnitude of the stresses, displacements etc will then be carried out, together with the accuracy expected from the experimental results.

- Preparation of either a strain-gauged or a photoelastic specimen for each group will then follow. The laboratory technician will check if the specimens are properly prepared and usable, if not, a new specimen will be prepared by him to be ready for the next session.

- Background reading materials on the use strain gauges, photoelastic techniques, and on the theory of finite element analysis will be made available in the laboratory for the students to consult. Demonstration on both experimental and numerical analysis for standard components will also be given.

- The students will be required to write up a laboratory diary during the three-hour session which will be read by the supervisor before the students leave the laboratory.

Session II involves:

- Loading the specimens and recording results.
- The necessary calculations and analysis of results will then be carried out. This will include uncertainty analysis, comparison with the estimations made in the previous session, and comparison with analytical solutions.
- A finite element analysis for the specimen will have been carried out by the laboratory supervisor or the demonstrator. The analysis will follow the location and type of loading chosen by the particular group of students previously. Whenever possible interactive mesh generation and stress analysis will be demonstrated. Printout of results, and graphical outputs will have been prepared in advance before the start of Session II. The students will have an opportunity to have hands-on experience with FE modelling.
- Again a laboratory diary will be required.

Session III includes:

- The introduction to the concept of failure criteria for a component. Experiments on determining fracture toughness, impact strength etc will be demonstrated. Students will then attempt to analyze the possibility for their chosen component to fail under certain loading and boundary conditions.
- Students who are reading for a materials engineering degree will spend more time on the fracture of both ductile and brittle materials, and their implication on design.
- An improved design for the component will be expected as a result of this analysis. The revised design of a particular group will be chosen for which its FE model will be modified by the supervisor for demonstration. The stress analysis which follows will illustrate the iterative nature of the design process.
- The results produced will be analyzed in detail and their implications discussed.
- Further demonstration on experiments such as plastic bending, unsymmetrical bending, shear centre etc will also be given if time allows.
- The students are also required to submit a laboratory diary for the session.

- A formal report will be submitted by selected project groups as part of their Part II laboratory assessment.

Stress Analysis - Part II

The formal introduction to the finite element methods will take place at the end of the second year of our three-year full-time course. The relatively free periods after the final examinations are being used for a thirty-hour one-week course in finite element modelling. Only students who have chosen stress analysis, or related subjects such as design, as one of their final year subjects will be allowed to attend the course, because of the limitation on both the number of workstations and tutorial staff support. Two $1^1/_2$ hour lectures, with $^1/_2$ hour break in between the lectures, will be given every morning for five consecutive days during the week. The afternoons will be used to practise using workstations and the interactive FE modelling package available at Newcastle. The concepts of both pre- and post-processing will also be introduced. The modelling will be carried out on standard benchmark problems, case study materials provided by selected final year dissertations, and both MPhil and PhD projects, and certain consultancy work undertaken previously at Newcastle. Current postgraduate research students will also be able to bring their own projects along for demonstration. A careful selection of these case studies will cover important concepts such as fatigue, fracture, failure mode effect and critically analysis, and optimum design. Some typical examples of case studies are presented in Appendix 1.

In fact, it is the author's submission that there is a need to generate a "finite element culture" right at the beginning of and also during the second year of an undergraduate course, so that the role which it can play as a computational technique in areas other than stresses, such as dynamic analysis, heat transfer, electric and magnetic potential, fluid flow, etc can and will be appreciated at an early stage of a student's university career. If the subject were only introduced in the third year, it would be too late for the method to be efficiently utilized in his final year undergraduate project work.

Stress Analysis - Part III

Equipped with a basic knowledge of the FEM and an interactive FE modelling package, a student will be able to undertake a stress analysis mini-project as part of the requirements of his/her final year examinations. Again a component will be allocated to each student for which a stress analysis by using both experimental and numerical methods, and also analytic methods if required, will be carried out. Similar approach as their Part II strength of material laboratory work will be adopted. The difference now of course is that each student will be working on their own, and all the experiment, modelling, and analysis will have to be undertaken by him/herself without the unlimited assistance from the laboratory supervisor. The students will have two terms to complete their investigations. There are eighteen Sun SPARC IPC workstations being available 24 hours a day for project work at the University Computing Laboratory. The operation of the testing machines in our Department will still be performed by the laboratory technical staff. Thus the stress analysis course at Part III level will consist of both a mini-project and a series of lectures on elasticity, stress functions, three dimensional stress analysis, problems on plane stress and plane strain, thermal stresses, and plate bending etc. The assessment of the former will be by dissertation (25% of the subject marks), and the latter by a two-hour examination (75% of the marks). The students are expected to attend a weekly tutorial of at least one hour duration in order to discuss their projects with the supervisor. A series of five introductory lectures on both finite difference and boundary element methods will also be given at the beginning of the Michaelmas term so that some of these methods can be applied to their projects if required. An interactive boundary element package is also available at the University of Newcastle upon Tyne.

Discussion

The successful completion of both the Part II laboratory and the Part III mini-projects depends to a very large extend on the careful preparations undertaken beforehand. The selection of specimens, and case studies has to take into consideration the resources available. Interactive graphics modelling is essential for a truly integrated approach. This of course would be best carried out on workstations, preferably with accelerated graphics facilities. The support from both postgraduate demonstrators and the department's technical staff is a key to the efficient teaching and running of the laboratory for such a course.

The students will present their findings and analyses in a dissertation which will be assessed in terms of their understanding of the underlying engineering and scientific principles involved with the projects, and their application and industry in getting to grips with the facilities and equipment available in the laboratories. Extra credit will be given to those who can show an appreciation of the wider implications of their projects in relation to design, manufacturing engineering, and to the quality assurance, failure analysis and the product liability aspects of their projects. Both the finite element analysis and experimental mechanics are notoriously difficult to assess, especially at the undergraduate level and also as a part of a course on stress analysis. A formal examination question can be set to assess the students' understanding of certain mathematical or experimental technique. But the more important aspects of the subjects in terms of the ability to design an appropriate model, or the ability to obtain accurate test results can hardly be examined formally in an examination hall and with a very restricted time scale. The assessment by dissertation will overcome these inherent difficulties.

The value of case studies when used for the teaching of engineering subjects has been universally recognized. From cases as grand a scale as, say, the space shuttle [12, 13] to relatively much simpler consultancy or final year design projects (e.g. Jenkins & Calder [14]), can be used to illustrate certain specific scientific and engineering principles, and thus can serve a particularly useful purpose. A major advantage which the students can gain from a case study is of course the availability of a self-contained report and usually together with the design of a component properly made, or with an experimental rig for the students to make an easy and speedy start. The professionalism and practical experience of the supervisors or laboratory assistants can come across more effectively when working with a well designed case study. Graduate students may also be able to bring with them their experimental rigs etc which can then be analyzed by undergraduates as an integral part of the learning process. If the case history of a project together with the difficulties encountered are to be understood, and the approach adopted in solving these difficult problems are to be fully appreciated, the students should be given all the available facilities at his disposal. The integrated approach presented in this section should help prepare the students to carry out their investigations effectively.

STRESS ANALYSIS, DESIGN, QUALITY ASSURANCE, AND PRODUCT LIABILITY

There has been a growing need for the introduction of the FEM at an early stage of design teaching especially with the continually decreasing prices for both computer hardware and software. Clarke [15] proposes that finite element analysis should be made available to designers at the concept stage. Holt and Radcliffe [16] suggest ways of integrating computers into design courses. Muster and Mistree [17] have also pointed out that engineering design is moving from an art towards a science, and discussed its impact on the education process. The finite element methods when properly utilized can indeed become a very powerful design tool. Simulated tests can be carried out well before the prototype product has been made. Re-designing a component will only require a re-meshing of the structure. With a good pre- and post-processor, this can be easily done. It is also relevant to point out that both the FEM and BEM would be indispensable for work on optimum design, although a detail discussion of which is beyond the scope of this paper.

The impending arrival of the European Single Market and the recent changes in the product liability laws embodied in the Consumer Protection Acts 1987 (see, for example Wright [18], and Brown [19]), require curriculum designers, especially those involved with stress analysis and design courses, to pay due attention to the requirements of the British, International, and European standards BS5750, ISO9000, and EN29000 respectively. Since all these standards on quality systems basically cover the same ground, only certain relevant aspects of BS5750 which concern both design and product liability will be discussed here.

BS5750 are a set of guidelines for companies to set up quality systems for the effective monitoring of the quality of their products. A company will be required to state their quality policies, to devise procedures for implementing these policies, and to provide instructions and specifications for the company employees to carry out these procedures. The appropriate procedures for corrective actions will also have to be designed so that any non-conformance to standard practices can be addressed. BS5750 are published in three major parts, i.e. Parts 1, 2, and 3; with Part 0 (in two sections) to provide guidance for the selection and use, and Part 4 to help with the interpretation of Parts 1, 2, and 3, as summarized in Appendix 2. Interested readers may also find Haldane [20] useful for an understanding of the practicalities of implementing BS5750 Part 1.

BS5750:Part 1 - **Specification for Design, Manufacture and Installation** states that a manufacturer should:

establish, document and maintain an effective and economical quality system to ensure and demonstrate that materiel[1] or services conform to the specified requirements. The documented quality system shall include quality management objectives, policies, organisation and procedures to demonstrate compliance with the requirements of this standard.

A company can register with the British Standard Institute as a firm of assessed quality capability. Following the above excerpt, if the company's activities include the design function, it should register according to the guidelines given in BS 5750:Part 1:1987. Although the courts will treat each case individually according to its own merits, the fact that a firm has taken the trouble to go through the registration process may constitute a case where the "**due diligence defence**" against criminal liability under Part II of the Consumer Protection Act 1987 (see Wright [18]). With regard to Part I of the Act, **strict liability** (see Ashley [21], Wright [18]) will be imposed if a **defect** in a product can be identified, and if it can be proved that such defect has caused certain injuries or damages. This means that it is not necessary to prove that there is negligence due to the **producer** of the product before compensation for damages can be claimed. Again the values of the compensation paid out to the injured party may be reduced if it can be demonstrated that a firm has taken all the necessary steps in order to ensure that its quality assurance management system has been properly designed and implemented. **Design review** should be regularly carried out to investigate the reliability of the products. Failure mode, effect, and criticality analysis (**FMECA**) should be an integral part of the review process (see e.g. Oakland [22]). This will ensure that every conceivable potential failure of a design has been considered. Here, the finite element technique should play a very important role. Potential failures of the product at each of the conceptual, embodiment, and detail design stages (Pahl and Beitz [23]) of the design process have to be analyzed. Tests will have to be carried out on modified designs (see e.g. Ashley [24]), to be followed by another finite element analysis, and so on. **Postgraduates** reading for their MPhil and PhD degrees at the Department of Mechanical, Materials, and Manufacturing Engineering are urged to attend lectures on the Quality Assurance and Product Liability modules of the MSc degree course in Quality Engineering currently offered at Newcastle. The importance of an integrated approach to tackling analytical, experimental, and numerical stress analysis cannot be over-emphasized.

---

[1] *Note the word **materiel** which is defined in BS 4778 as "equipment, stores, supplies and spares that form the subject of a contract."*

The emergence of the concept of **Simultaneous Engineering (SE)** serves to illustrate further the need for an integrated approach to modern engineering practice. Simultaneous Engineering has been defined by Hartley [25] as:

> *"the use of a multi-disciplinary team to control a design project with the use of certain specific techniques."* *Of these techniques, Quality Function Deployment (QFD) and Design for Manufacture and Assembly (DFMA) are the most important.*

Hartley [25] explains that the project is led by a Task Force consisting of people from Product Design, Manufacturing, Marketing, Purchasing, Finance, and Principal suppliers of manufacturing equipment and components. These people will "remain in the Task Force at least until volume production is reached, and sometimes until the model is superseded."

The main advantage of this approach is that several departments can interrelate with each other and thus can work **simultaneously.** He reports that "SE has been shown to work by companies such as Volkswagen and Rover, which have cut lead time from 48 to 36-40 months and from 72-80 months to 48 months respectively." One of the techniques which can be used under Quality Function Deployment is to send the Task Force to visit customers, at the beginning of the project before work has started, to survey their views on aspects of the design. The findings are translated into an engineering specifications which can then be judged in relation to the competitors products. Product Design shares the same computer database with, say, the tooling department as well as other departments, via a CAD/CAM route. Thus tooling can be made simultaneously with the execution of other activities once a design has been finalized, so that the lead time for the fabrication or manufacture of tooling could be drastically reduced.

CONCLUSIONS AND RECOMMENDATIONS

The need for an integrated approach to the modern product development process has promoted a radical rethinking of both the undergraduate and postgraduate curriculum. Numerical analysis techniques such as both the finite element and boundary element methods play a very important role in ensuring the basic design of a product is sound at every stage of its development. The teaching of the traditional Strength of Materials laboratory should be integrated with the teaching of the finite element analysis techniques. A coherent undergraduate course structure has been proposed. The important contribution of properly selected case studies should be recognized. Assessment of student performance can be based on the submission of a dissertation on a mini-project. The project will contain analytical, experimental, as well as numerical work. Major limitations to adopting such an approach will be the resources available to a Department or an Institution.

Lecturing periods should be devoted to the introduction of the implications of the failure of a component with regard to product liability and the Consumer Protection Acts 1987, in the context of the United Kingdom's involvement with the European Single Market.

The fact that an integrated approach to stress and design analysis would allow a firm to benefit from the adoption of modern concepts such as Simultaneous Engineering should also be emphasized.

Gazing into the future, further developments in the combination of FEM and traditional experimental techniques such as photoelasticity, and Moiré interferometry; the developments in the hybrid experimental-computational approach for the photomechanics techniques, together with the availability of the ever more powerful computing hardware, the finite element method will surely gain even more prominence in the product design and development cycle in the engineering manufacture and service industries.

ACKNOWLEDGEMENTS

The author wishes to thank Professor P.M. Braiden most sincerely for his invaluable guidance and numerous discussions with the author during the preparation of this paper.

REFERENCES

[1]   **Holman, J.P.**, *Experimental Methods for Engineers*, Fifth Edition, McGraw-Hill, **1989**, ISBN 0-07-029622-7.

[2]   **Adeli, H.** and **Chen, Y.**, Micro-computer graphics in teaching mechanics of materials, *International Journal of Applied Engineering Education*, Vol. **5**, No. 4, pp 471-476, **1989**.

[3]   **Potts, J.K.** and **Oler, J.W.**, *Finite Element Applications with Micro-computers*, Prentice Hall, **1989**, ISBN 0-13-317439-5.

[4]   **Roth, R.**, On constructing free-body diagrams, *International Journal of Applied Engineering Education*, Vol. **5**, No. 5, pp 565-570, **1989**.

[5]   **Dixon, J.R.**, New goals for engineering education, *Mechanical Engineering*, pp 56-62, March **1991**.

[6]   **Berghaus, D.G.**, Combining photoelasticity and finite element methods for stress analysis using least squares, *Experimental Mechanics*, Vol. **31** (March 1991), No. 2, pp 130-134, March **1991**.

[7]   **Morton, J., Post, D., Han, B.**, and **Tsai, M.Y.**, A localized hybrid method of stress analysis: a combination of Moiré interferometry and FEM, *Experimental Mechanics*, Vol. **30**, No. 2, pp 195-200, June **1990**.

[8]   **McKelvie, J.**, Full-field methods in experimental mechanics: a perspective and prospect, *Applied Solid Mechanics - 3*, pp 222-243, Elsevier Applied Science, **1989**, ISBN 1-85166-435-1.

[9]   **Blicblau, A.S.**, The modern engineering curriculum in materials, *International Journal of Applied Engineering Education*, Vol. **7**, No. 1, pp 4-8, **1991**.

[10]   **Miniatt, E.C., Waas, A.M.** and **Anderson, W.J.**, An experimental study of stress singularities at a sharp corner in a contact problem, *Experimental Mechanics*, Vol. **30**, No. 3, pp 281-287, September **1990**.

[11]   **Stanley, P.**, Application of the thermoelastic effect to stress analysis, *Applied Solid Mechanics - 3*, pp 269-286, Elsevier Applied Science, **1989**, ISBN 1-85166-435-1.

[12]   **Looft, F.J., Labonté, R.C.** and **Durgin, W.D.**, The evolution of the WPI Advanced Space Design Program - an evolving program of technical and social analysis using the NASA space shuttle for engineering education, *IEEE Transactions on Education*, Vol. **34**, No. 1, pp 20-26, February **1991**.

[13]   **Bonsall, C.A.,** The NUSAT I project - government, industry, and academia learning together, *IEEE Transactions on Education,* Vol. **34,** No. 1, pp 15-19, February **1991.**

[14]   **Jenkins, C.** and **Calder, C.A.,** Transient analysis of a tennis racket using PC-based finite elements and experimental techniques, *Experimental Mechanics,* Vol. **30,** No. 2, pp 130-134, June **1990.**

[15]   **Clarke, R.B.,** Making finite element analysis available to designers at the concept stage, *Proceedings of the Institution of Mechanical Engineers, Part B,* Vol. **202,** No. B2, pp 81-86, **1989.**

[16]   **Holt, J.E.** and **Radcliffe, D.F.,** Some perspectives for integrating computers into design courses, *International Journal of Applied Engineering Education,* Vol. **7,** No. 1, pp 31-34, **1989.**

[17]   **Muster, D.** and **Mistree, F.,** Engineering design as it moves from an art towards a science: its impact on the education process, *International Journal of Applied Engineering Education,* Vol. **5,** No. 2, pp 239-246, **1989.**

[18]   **Wright, J.C.,** *Product Liability - the law and its implications for risk management,* Blackstone Press Ltd., **1989,** ISBN 1-85431-036-4.

[19]   **Brown, S.,** *The Product Liability Handbook: Prevention, Risk, Consequence and Forensics of Product Failure,* New York: Van Nostrand Reinhold, **1991.**

[20]   **Haldane, T.,** *Meeting Quality Standards: A Practical Guide to BS5750 Part 1:1987,* Pergamon Press, Oxford, **1989,** ISBN 0-08037-112-4.

[21]   **Ashley, S.,** Confronting product liability, *Mechanical Engineering,* pp 46-47, March **1991.**

[22]   **Oakland, J.S.,** *Total Quality Management,* Heinemann, **1989,** ISBN 0-434-91479-7.

[23]   **Pahl, G.** and **Beitz, W.,** *Engineering Design,* The Design Council, **1984,** ISBN 0-85072-124-5.

[24]   **Ashley, S.,** Safer products through tough testing, *Mechanical Engineering,* pp 63-65, January **1991.**

[25]   **Hartley, J.,** Simultaneous Engineering - key to better design and manufacture, *Automotive Engineer,* April/May 1991, pp 50-52.

APPENDIX 1

Some examples of case studies are presented below:

- Optimum design of porcelain switchgear insulators.

- Seismic modelling of live tank high voltage circuit breaker.

- Thermal analysis of an umbilical winch for subsea pipe and cable burial equipment.

- Simulation of crack propagation in plates and shells.

- Propagation of collinear inclined cracks in brittle materials.

- Crack propagation for two equal skew-parallel cracks in thin plates.

- Simulation of entry and exit failure during intermittent metal cutting.

- An investigation into the characteristics of constant, linear and quadratic boundary elements.

- A direct boundary element method for transient thermoelasticity problems.

- 3-D elasticity analysis of a low pressure turbine blade root by the boundary element method.

APPENDIX 2

British Standard Specifications BS 5750 on Quality Systems:

- **Part 0 - Principal concepts and applications**
  BS 5750:Part 0:Section 0.1:1987 - Guide to selection and use
  BS 5750:Part 0:Section 0.2:1987 - Guide to quality management and quality system
    elements

- **BS 5750:Part 1:1987** - Specification for design/development, production, installation
    and servicing.

- **BS 5750:Part 2:1987** - Specification for production and installation.

- **BS 5750:Part 3:1987** - Specification for final inspection and test.

- **BS 5750:Part 4:1990** - Guide to the use of BS 5750:Parts 1, 2, 3 (Formerly BS
    5750: Parts 4, 5, 6).

INTEGRATION OF A 'HANDS-ON' FINITE ELEMENT COURSE WITH A TRADITIONAL MECHANICS COURSE

D.J. GRIEVE
G. ROBERTS
DEPARTMENT OF MECHANICAL ENGINEERING
POLYTECHNIC SOUTH WEST
PLYMOUTH

ABSTRACT

This paper considers the recent development of a final year option in Finite Element Analysis where the emphasis is on a hands-on project type approach. An extension of this approach is proposed where the projects form the basis of the tutorial work associated with a more traditional final year mechanics course. A series of such projects are suggested and these are also linked to experimental investigations.

Keyword. Mechanics, Stress, Vibrations, Finite Element, Case Studies.

INTRODUCTION

The development of mechanical engineering degree courses in the UK over the period 1900-1960 followed a fairly well defined curriculum where emphasis was on analysis in the subject areas broadly classified as Theory of Machines, Strength of Materials, Applied Thermodynamics and Fluid Mechanics. These were usually backed up with courses in Mathematics, Electrical Engineering and Metallurgy.

In the period 1960-1991 the increase in subject matter incorporated into undergraduate courses has been alarming and this has placed immense pressure on the classical analytic content. Typically additional subjects include : electronics, instrumentation, control theory, computer programming and interfacing, material science, statistics, design, production technology and control, business studies, project work, numerical methods, CAD and CAM packages and engineering application exercises. Most courses attempt to include as much material as possible by offering a wide range of options in second and third years. This however causes programming difficulties and has resulted in wide gaps in students knowledge of basic concepts.

At Plymouth this problem was addressed during the period 1975-1990 when the discipline of obtaining CNAA approval for courses resulted in much good debate on curriculum development.

One early change involved the combination of the first year machines, stress and fluids subjects into an integrated course covering rigid body, deformable body and fluid statics and dynamics.

The next stage which is reported in this paper is the extension of the integrating exercise into years 2 and 3 whilst also making space for the inclusion of FEA.

MECHANICS II

The specification for this course was that it should cover the basic subject matter in stress analysis and system dynamics with which, it was felt, every student should be familiar. Fluids was combined at this stage with thermodynamics. In this context Mechanics II is a compulsory mechanical engineering subject.

The syllabus tends towards stand alone topics although continuity with the Mechanics I course is ensured by involving the same staff with each year.

One of these topics is an introduction to FEA. The approach is very traditional building from matrix methods in simple structural analysis to continuum analysis using constant strain triangular elements. Emphasis is on element stiffness derivations, structural stiffness assembly, boundary condition partioning and solution by direct matrix inversion. Tutorial work includes solution of simple problems using an 'in-house' suite of programmes originally written for and running on BBC 48K machines but recently converted to run on the 'new' Opus 386 SX networked system. Graphical output is available to give more of a real feel to the package but the emphasis is on inspection of matrices at all stages and the development of an understanding of the basic principles of the method.

MECHANICS III

As stated above, the Mechanics II course covers what is believed to be essential knowledge in stress analysis and dynamics. It follows that Mechanics III forms one of a series of final year options, this could cause problems because, if the numbers opting for the course were greater than ~20 serious problems would be experienced with work station availability. This has not happened so far, as the grapevine information on the required student commitment precludes the less diligent!

Initially the Mechanics III course was conceived as essentially a hands-on FE application course where a series of mini-projects were tackled within the broad framework of a lecture course in FE methods. The somewhat misleading choice of title was political. After running this course for two years [1,2], one major problem experienced has been that discussion of the FE projects is seriously hampered by the student's limited awareness of classical analytical methods. Although students are advised to consult standard texts the absence of any structured reading makes this a very time-consuming and inefficient approach.

It has recently been concluded that it should be possible to structure the projects in such a way that they can be used to integrate lecture courses in stress and dynamic analysis, finite element theory and experimental methods.

The suggested approach really involves the use of 'hands-on' project work to replace the traditional numerical tutorial work. The flexibility and range of 'examples' being much more real and informative.

A simple example is the problem of a hole in a tension strip. Students working in pairs model the problem on one of the 16 workstations linked to a dedicated PRIME 9755 and running the PAFEC suite. Various geometrics, element types and meshes can be investigated. In parallel with this 'hands-on' work the lecture course covers:

(a)   An introduction to two dimensional theory of elasticity leading to the classical Kirsch solution for a hole in an infinite plate, [3].

(b)   An introduction to transmission photoelasticity leading to an investigation involving material stress/fringe calibration and isochromatic recording of shear stress distribution.

(c)   A discussion of element types and the shape function approach, meshing techniques, boundary conditions for problem reduction by symmetry and mesh density refining for accuracy.

THE SELECTION OF PROJECTS

Associated with the change of emphasis in the project work described above is the need to cover a reasonably structured mechanics course. Such a course might include: (a) Stress Analysis - two dimensional elasticity, plate bending, St Venant torsion theory, axisymmetric problems, plasticity; (b) Dynamics - analysis of one, several and multiple degree of freedom systems to investigate natural frequencies and response to harmonic or general transient loading.

Table 1 gives suggested projects and which areas they may integrate.

GENERAL COMMENTS

The following discussion is mainly based on the experience already gained in running the existing project based course. A more critical discussion of the proposed integrated course will be relevant only after sufficient experience has been accumulated.

Student Response
For academics versed in the traditional 'lecture/tutorial/theory demonstration experimental' approach the student response and staff-student interaction is exciting and rewarding albeit time-consuming. In contrast to cajoling students to solve tutorial examples we have had to curtail their enthusiasm for fear of adverse effects on other subjects. Demand for extended availability of work-stations in evenings and weekends symbolises this phenomenon.

Assessment
The subject is assessed mainly by marking the project reports. Students generally work in pairs the grouping being changed for each project, individual contributions are highlighted. One project is an individual exercise.

The subject counts equally with other examined subjects and hence good assessment is important. In general the report presentations are very professional but significant differences in detailed analytical discussion and critical use of FE models makes ranking relatively easy. As with most continually assessed courses a conscious effort has to be made to bear in mind that 70% is a first class honours mark. It is also noticeable that standard deviations are often lower than traditional examined subjects because poor performance in one or two assessments can be 'clawed back'.

Extension to Other Areas
The availability of PAFEC modules in heat-transfer and composite material elements offer opportunities for integration in other areas over the range of degree courses offered in the department viz, Mechanical, Systems, Composite and Manufacturing Systems Engineering.

Resource Implications
The conflict over investment between main-frame and micro-networks presents problems when significant commitment has been made in an existing CAE suite. Currently the PAFEC micro package is being evaluated within the department but this will not run on 386 SX network.

A MSCPAL package has also been used successfully on some projects using 286 machines and discussions are currently underway to implement this on the network - this would enable larger student groups to be accommodated on the option.

Teaching Methods
In all three years of the course prepared typewritten sets of notes are made available to the students at the cost of reproduction ~£5-£10/year. This allows a directed learning approach where lectures highlight important aspects whilst detailed algebra and worked examples can be studied in private. At first and second year level this releases significant time for students to cover the tutorial work and at third year to carry out the FE projects. The first year package is shortly to be published commercially [4].

CONCLUSION

The existing FE course has proved popular and academically successful. It is hoped that the modifications discussed in this paper will broaden the effectiveness from a specialised FE course to a truly integrated final year mechanics course.

TABLE I
A selection of projects and associated subject matter

Project Discussion	Associated Theory [5]	Experimental Work	FE Considerations
Torsion of solid and thin walled sections	St Venant's torsion analysis, shear flow methods	Strain gauge and deformation measurements on a square section solid shaft and on a 3-cell aerofoil section	Application of brick elements. Direct solution of Poisson's equations. Problem of boundary conditions for thin walled structures
Transverse loading of a rectangular plate	Plate theory	Strain gauge and deflection measurements on plates of various thicknesses	Shell elements, membrane and bending effects. Boundary problems of SS plates where the corners rise
Analysis of a rotating axial flow gas turbine disc	Axisymmetric problems in two dimensional stress analysis with body forces	Strain gauge measurements on rotating systems. Use of slip rings	Use of symmetry to reduce problem size. Consideration of body forces
Cyclic plastic strain amplitude in areas of stress concentration	Yield criteria and Prandtl-Reuss equations to define elasto-plastic problems. Cyclic tension - compression hysterisis curves. Application in low-life fatigue analysis [6]	Local strain measurements at a notch over a loading cycle. Comparison with Neuber's Rule	Plasticity solution techniques. Material behaviour representation
Dynamic analysis of a simplified vehicle suspension system	Vibration analysis of single degree of freedom systems. Use of lumped masses to avoid coupling. Various loading response	Experimental investigation on a Westward Garden Tractor using an electro-mechanical vibrator along with B and K vibration analyser	Dynamic analysis of lumped mass systems with spring and damper elements
Dynamic analysis of a 'two-storey' building model	Vibration response a two degree of freedom system subjected to harmonic excitation	Stroboscopic observation of resonance modes using rotating eccentric mass excitation of the model	Use of beam elements - axes orientation. Structural damping methods. Model and frequency response solutions

TABLE I

Project Discussion	Associated Theory [5]	Experimental Work	FE Considerations
Dynamic analysis of a free-free beam	Continuous system. The beam transverse vibration theory. 'Exact' solutions of the free-free beam	Vibration analysis using: (a) Electro mechanical vibrator and white noise; (b) Impact hammer	Modal analysis to determine the effect of different damping models. Use of different numbers and location of master nodes

REFERENCES

1. Roberts, G. and Grieve, D.J., Teaching of finite element analysis and mechanics in a BEng mechanical engineering course. NAFEMS Benchmark, October 1990.

2. Roberts, G. and Grieve, D.J., A case study approach to teaching mechanics to final year mechanical engineering students. Innovative Teaching in Engineering, Sheffield, September 1991.

3. Timoshenko, S.P. and Goodlier, J.N., Theory of Elasticity, McGraw-Hill.

4. Roberts, G. and Gates, B.P., Introduction to mechanics of rigid bodies, deformable bodies and fluids for engineers. (To be published).

5. Thomson, W.T., Theory of vibration with applications. 2nd Edition, Allen and Unwin.

6. Socie, D.F., Estimating fatigue crack initiation lives in notched plates under variable loading histories. T. and A.M. Report No. 417. University of Illinois, 1977.

BETTER THAN GUESSING THE STRESSING

APPLICATIONS OF THE FINITE ELEMENT METHOD IN AN UNDERGRADUATE COURSE

D F S Middleton
Manufacturing Engineering
University of Dundee
Dundee   Scotland

ABSTRACT

In Engineering Design courses the calculation of stress and strain is traditionally done by pencil and paper methods. Rather gross and sometimes dubious assumptions are made regarding stress distribution and concentration and mainly in a one dimensional manner. Deflection analysis for realistic designs of component are not handled any better and usually involved graphical analysis. Computer based solutions were available but rarely found application in undergraduate courses. The advent of automatic finite element calculation software eg PAFEC allowed the modelling of components to achieve solutions which could be obtained in reasonable time, in variety and with greater credibility than previous methodology allowed. At the University of Dundee the first undergraduate experiences were in the honours year projects where the time and enthusiasm was there to break new analysis ground. Third year students working on design-and-make projects were a natural next level to adopt the method and this has remained the first encounter point with a professional level system to date. The basics of FE analysis are encountered however at second year mechanics of solids level. Staff response to FE software was patchy and sometimes hostile especially by the purists who had to see something written out in longhand detail. The FE experience has been a liberating one freeing the mind from the the rigours and methodology of analysis and allowing more attention to be given to engineering the solution. The early days of large numerical input which led to even larger output are now behind us and graphical aids in modelling and results presentation are available. Mesh design and development is an area where we still have much to learn.

INTRODUCTION

The traditional Mechanical Engineering course had a considerable emphasis on mechanics of solids problem solving. What follows here should be read in the context of teaching and learning in such an undergraduate engineering course.

Stress and deflection analysis played a major role forming the main size determination activity. If the product comprised beams of constant cross section without any unfortunate deviations such as holes notches or similar discontinuities then standard solids mechanics solutions were available.

However most mechanical engineering products are not like that and in reality are chunky boxy things with all manner of geometric complexity and discontinuity. If stress calculations become difficult with geometric complexity then deflection calculations are even more difficult and time consuming.

To extend pencil and paper analysis involved in problem solving to include such topics as heat transfer and dynamic analysis of real shapes was almost impossible beyond a one dimensional treatment. Deep study in an honours course may just have allowed some progress to be possible on simplified structures.

In this context I am am reminded of papers at a recent advanced dynamics research conference where most of the titles on classical dynamics research were applied to simple beams and cantilevers.

In short the received wisdom of having to solve engineering problems by classical methods constituted a serious limitation and constraint to the solution capability of the engineering design lecturer and in consequence to the engineering student.

The late 1970's saw the availability of the finite element method in a form readily applied to engineering problems and the scene could begin to change I think for the better.

Early Experiences

The first manifestations of finite element analysis as I saw them were a lecture by Zienkiewicz in the early 1970's and a paper given by GEC design engineers at a seminar in Dundee. As the cognoscenti will know the first experience was a privelige to hear one of the fathers of finite element analysis but the fact was very few of us knew what the lecture was about or how it could be applied.

Remember that at this time there were little or no commercial finite element software packages available indeed the term software package was not even known. There would however have been programs available in research labs and stress offices of major companies.

The GEC paper which concentrated on the determination of stresses in the proximity of turbine casing nozzles struck a note of interest however.

Here was a real product of considerable geometric complexity being stress analysed with useful results.

Now this problem had been formulated and solved by numerical methods specialists using the finite element method with in house software and ample computing power but the speaker was an engineer and one could follow the engineering of the solution, eg the changes necessary to improve the design without getting bogged down in the solution method.

I could not help thinking that it had to be an industrially based engineer who presented the finite element method in this product orientated way as an academic would have gloried in the mathematics of the solution method and lost sight of the engineering problem. This attitude is reflected in the Higher Education of our young people where respectibility for engineering is sought under the cloak of mathematical complexity.

The exception usually proves the rule and in this case in the person of a lecturer at Nottingham University named Richard Henshell. His research programme at Nottingham led to an automatic finite element package which was marketed in 1975 by his then University based company PAFEC.

The advent of this facility opened the gates to new horizons of solutions capability for engineers. Colleges and Universities could have the licence to use the package at low cost (about £200 per annum) because of the fact that much of the research and development had been supported by the then Science Research Council.

PAFEC also ran economic training courses for intending and improving users and attendance at one of these courses in 1977 marked my introduction to using the finite element method. There were by this time other finite element analysis programs available eg ASAS and NASTRAN but the cost of using these was prohibitive.

Breaking the Ice

The plan in my Design teaching is the use finite element analysis as a tool to achieve an end rather than form an object of study in itself and I believe that most industrially based designers take a similar view.   The theory behind finite element analysis is given in the Solid Mechanics Course as part of the study of Theory of Elasticity.

It may be useful at this point to comment on the attitudes of some colleagues to the proposed use of finite element analysis in undergraduate courses.   One objection commonly voiced was that the students would not learn much about structured analysis if they did not personally write out the strain energy equations, set up the matrices and solve the problem long hand.   There was a fear that the finite element solution package was a black box not to be tinkered with.   To the designer of course this attitude was nonsense, the designer is interested in specifying a good reliable product to an acceptable level of accuracy and the analysis is a necessary evil to ensure that accuracy.

This was the entry to a cultural change, a change in attitude which requires the engineer, academic or otherwise, to place his trust in the analytical work of others.   It was necessary to accept that if the data file was correct the finite element package would calculate the results reliably. It seems commonplace now to accept many kinds of complex problem solving packages but twelve years ago there was much scepticism.   There should be caution of course, some solution processes are less accurate or reliable than other, errors can be made by the package programmers, but this is where agencies such as NAFEMS play a most important role with their regular survey reports of tests carried out on the leading finite element packages worldwide. These are reported in the excellent NAFEMS periodical "Benchmark".

Getting Started

The method of operation for most finite element solution packages is to prepare a data file as stage one of the process.

Figure 1 shows the data file for a PAFEC analysis of the hole in plate example.

In this file the geometry of the problem is specified by co-ordinates of nodes together with element type, material properties, loading and a statement of the type of solution required. Although much of the early work in finite elements was done in structural analysis for stress and deflection current practice covers many problem solution fields such as heat transfer, fluid flow and dynamics.

Getting the data file to an operational state should be accepted as a process of development. If the student can accept this it helps to overcome frustration of failure on early attempts. In the undergraduate teaching context a well designed finite element package is a learning tool if the student is prepared for the experience. In the PAFEC software which comprises ten solution stages, of which the data file forms stage one, the process can be represent levels of solution and each solution stage will provide a report on the progress of solution. This means the student can step through the process amending his data file after each run as the reports are produced. So feedback is provided by the package which forms a kind of "self learning process" - computer assisted learning if you like. Clarity of error messages is very important here and interpretation of these messages causes much of the difficulty in the early stages.

First trials
When the staff members involved had achieved a working competence in the art of running the finite element package it was time to try it out on the students.

The obvious place to take first steps was with an honours project student. There was much shaking of heads at the thought of introducing such a topic for an honours project but it was done.

The student I got was not "one of our most able students mathematically" but he did have an interest in design and like most of our students came to University to study engineering not mathematical analysis.

```
 1     CONTROL
 2     NAME.PLA1
 3     PHASE=3,5,8,10
 4     CONTROL.END
 5     NODES
 6     NODE.NUMBER,X,Y
 7     1      0       0
 8     2      .05     0
 9     3      .08     0
10     4      .125    0
11     5      .31     0
12     6      .5      0
13     7      .0447   .0223
14     8      .08     .0223
15     9      .03     .04
16    10      .08     .04
17    11      .125    .04
18    12      .31     .04
19    13      .5      .04
20    14      .01581  .04743
21    15      0       .05
22    16      0       .08
23    17      .03     .08
24    18      .08     .08
25    19      .125    .08
26    20      .31     .08
27    21      .5      .08
28    22      0       .125
29    23      .03     .125
30    24      .08     .125
31    25      .125    .125
32    26      .31     .125
33    27      .5      .125
34     ELEMENTS
35     ELEMENT.TYPE=36210
36     TOPOLOGY
37     2    3    9    10   0   7   8
38     3    4    10   11   0   8
39     4    5    11   12
40     6    13   5    12
41     15   9    16   17   14
42     9    10   17   18
43     10   11   18   19
44     11   12   19   20
45     13   21   12   20
46     16   17   22   23
47     17   18   23   24
48     18   19   24   25
49     19   20   25   26
50     21   27   20 26
51     RESTRAINTS
52     NODE.NUMBER,PLANE,DIRECTION
53         4          2          2
54         22         1          1
55     PLATES.AND.SHELLS
56     PLATE.NUMBER,MATERIAL.NUMBER,THICKNESS
57      1            1            .001
58     IN.DRAW
59     DRAWING.NUMBER,TYPE.NUMBER,INFORMATION.NUMBER

60          1          1          23
61     PRESSURE
62     PRESSURE,LIST.OF.NODES
63     -8.0E06    6   13   21   27
64     STRESS.ELEMENT
65     START,FINISH,STEP
66     1   228   1
67     OUT.DRAW

68     DRAWING.NUMBER,PLOT.TYPE

70          3              32
71     END.OF.DATA               0 ERRORS
```

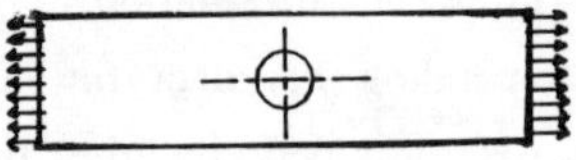

FIGURE 1

PAFEC DATA FILE

FOR HOLE IN PLATE ANALYSIS

The first problem he tackled was the hole in a rectangular plate under tension in one direction. This is a common and useful problem as a starter providing a two dimensional stress field for which known solutions exist based on classical elasticity theory.

Figure 2 shows a quarter of the model after solution with stress vectors plotted.

The first runs did not work due to syntax errors but the built in error detection warnings assisted in clearing these difficulties.

The specification of restraints is a major problem at the early stages and even at later stages of familiarisition. The development of this hole in plate problem proved to be a very educational exercise. While a two dimensional stress field was expected in the region of the hole it was surprising in early runs to find such stresses in way of the vice grips at the load application point where incorrect restraints had been specified. The wonders of this finite element analysis tool began to appear, one could experiment with different restraints and loading to observe the effects.

In classical stress analysis calculations one selects the point for examination as is also the case in strain gauge experimentation. In finite element work the whole field can be seen including the unexpected. This student proceeded to consider the more complex problem of stresses and of temperature distribution in a diesel engine piston. Students in subsequent honours years proceeded to experiment with the wealth of features available.

Figure 3 shows a hydraulic piston section in an axisymmetric model

These students have the time and inquisitive energy to pursue lines of enquiry and development which academic staff cannot equal dueto constraints of teaching and administration but it does need a bit of courage at the start to hand over this authority to the student. There is a tradition among some staff of honours student projects being based on known territory only and the fear of the new can be present.

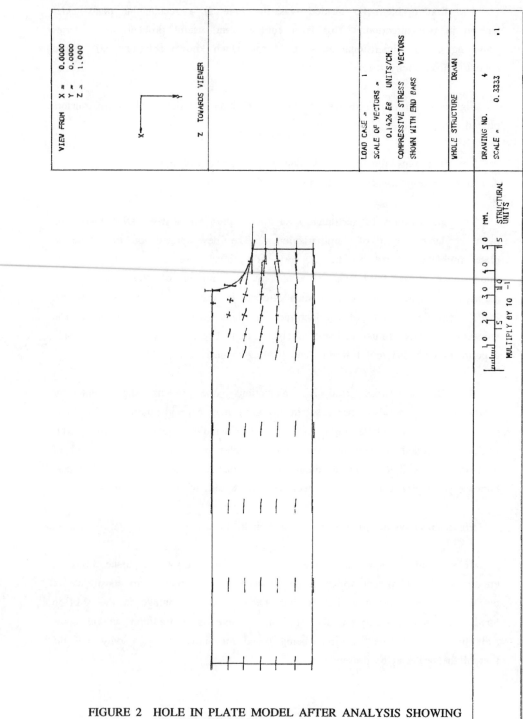

FIGURE 2   HOLE IN PLATE MODEL AFTER ANALYSIS SHOWING
STRESS VECTORS

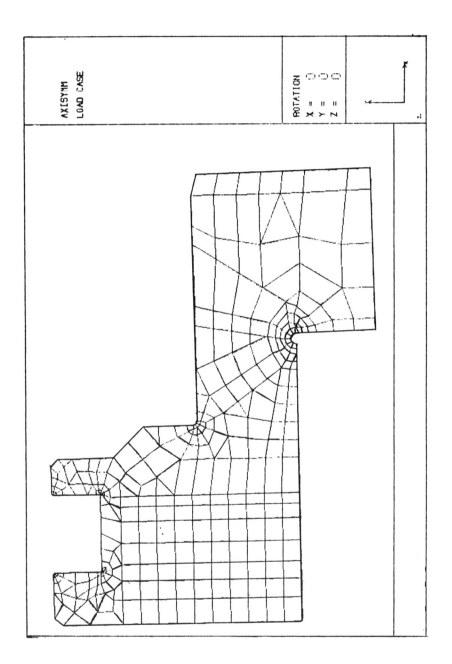

FIGURE 3   HYDRAULIC PISTON MODEL SHOWING DEFLECTED SHAPE.

It should be made clear that where possible an experimental rig was constructed to test a physical model of the part and also reference was made where appropriate to known exact classically determined solutions.

Introduction to earlier years

The Mechanical Engineering degree at Dundee is a four year honours course which is the traditional Scottish model. I sought an opportunity to introduce the finite element analysis in the third year.

As a lecturer in Mechanical Engineering Design I was setting project work on various topics and this seemed to be an activity suited to use of a finite element analysis package.

The project chosen was to be a design, make and test type of exercise based on a simple structure.

A specified gap was to be crossed and a load carried mid span on a structure which to make it interesting was to be of minimum mass consistent with carrying the load safely. Safety was defined as a limiting stress anywhere in the structure, namely a beam.

The project comprised design, analysis, manufacture and testing of the beam The material was rigid polystyrene foam of maximum dimensions $600 \times 100 \times 50$.

In order to keep the problem tractable and to limit the complexity of the models it was stated that the shape must by $2\frac{1}{2}$D ie an XY shape with the thickness in discrete steps. The method of deciding the shape need not concern us here but it was clearly an optimisation problem.

The design was then programmed for cutting on an NC milling machine which provided a manufacturing engineering involvement. The beam was then tested to determine its ability to carry the required load and the lightest beam to do so elastically was judged the winner. The mid point deflection was measured. Beams under test are shown in Figures 4 and 5.

This sort of challenge created an enjoyable environment in which to use finite element analysis. After designing for optimum mass distribution the

FIGURE 4   BEAM UNDER TEST SHOWING LATERAL BUCKLE

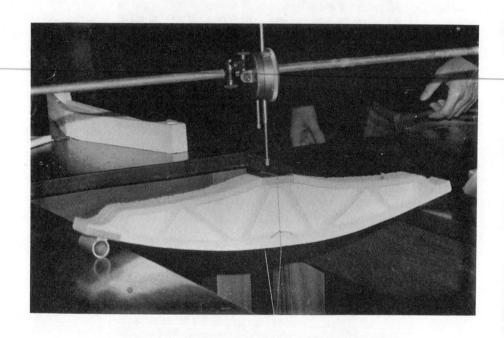

FIGURE 5   BEAM TEST SHOWING LOAD AND DEFLECTION
MEASUREMENT

design had to be modelled for stress and deflection analysis and shown to have a reasonable chance of success prior to manufacture.

The project was first run as an individual project which proved to involve rather too much demand on human and computing resources and also left the weak students struggling. Principal observations included :

The groups approach was much the preferable way for this work and a team spirit readily evolved.

The students rapidly learned the limitations on geometry of elements eg element corner angles too acute.

Discretisation of the model into a usable mesh proved a challenge.

Their course knowledge on stress concentration led to application of increased mesh density eg at corners and fillets.

It became apparent to the student that once the geometric model for deflection had been created minor additions to the data file could produce solutions for stress, natural frequency, modes of vibration and if required temperature distribution.

The only finite element result which could be readily verified from the simple test was deflection and reasonable agreement was experienced.

Parallel analyses by alternative pencil and paper graphical means were run by other team members for comparison. While the loading test was intended to measure the ability to carry the 2kg target load elastically every student carried the test to failure and several failed beams are shown in Figure 6.

Conclusion
I believe that it has been shown that there is a place for the use of finite element analysis in the undergraduate course. Once the student "gets started" the process provides a self learning environment.

Keeping the problem simple in the beginning pays dividends and the initial foray is best restricted to a simple rectangular cross section beam.

FIGURE 6   SET OF BEAMS AFTER LOADING TO FAILURE

It was interesting to note subsequent to the above project work the spread of finite element analysis applications to a range of problems in the honours year. I believe that the students once given a taste of the method were prepared to apply it to the problems they encountered in their honours projects even to the extent of introducing their supervising staff to its benefits.

My own view was that using the finite element method in its automatic readily applied form has been a liberating experience freeing one from the constraints of limited analyses or the need for specialised computing knowledge The growing crop of PC based packages will no doubt extend the application of the method to earlier years of the undergraduate course where one expects to find simple beam and cantilever models compared with solutions by classical analysis.

A PRACTICAL APPROACH TO TEACHING THE FINITE ELEMENT METHOD

MAHMOOD H DATOO
BSc PhD CEng MRAeS
Dept of Civil and Structural Engineering
South Bank Polytechnic
Wandsworth Road, London SW8 2JZ, UK

ABSTRACT

Teaching of the finite element method to the degree level
final year undergraduate students is accomplished
predominantly by simple case studies hands-on experience, and
supplemented by an overview of the theoretical analysis. The
output are compared with solutions obtained by the classical
structural analysis methods and thin-walled theory already
covered in detail in the structures syllabus. Students are
introduced to modelling techniques by a progressive usage of
line, membrane and shell elements. For the two dimensional
elements, the effect of mesh refinement is investigated by
studying the output of displacements and stresses. Students
are purposefully discouraged to use any automatic data
generation facilities so as to get the feel of the method.

INTRODUCTION

The increasing usage of the finite element method in industry
has meant that the graduates should preferably be acquainted
with the theoretical overview of the finite element method
and basic modelling techniques. The subject of the finite
element method is part of the syllabus in the Advanced
Structural Analysis unit of the final year undergraduate
degree course. The author felt that the best approach to
expose the students to the finite element method at an
introductory level would be a practical hands-on sessions,
rather than a purely classroom based theoretical aspects.

Hardware and software The choice of the hardware and software was made by default due to their availability on site. The Polytechnic has a number of Apollo workstations on site on which the PAFEC (Level 7.200) finite element package is mounted. Also available on the workstation is the PIGS pre and post- processor associated with the PAFEC system. Consequently, the PAFEC system was used. It was however emphasised to the students that whilst most commercial packages will differ from one another, the essentials of the data preparation and output would be similar. Moreover, the art of good modelling and results interpretation and verification is naturally independent of any package.

Classical Structural Analysis

In the final year of the undergraduate course, students opting for the Structural Engineering degree have to study the Theory of Structures and Advanced Structural Analysis units. The Theory of Structures course contains the subject of structural analysis by the matrix algebra methods. Here, the stiffness method of structural analysis is covered in detail in the context of plane truss and frame structures. The concepts of nodes, definition of global and local axes, nodal loads and restraints, element stiffness, assemblage of element stiffnesses, methods of solutions, would thus already be familiar to the student before starting the finite element sessions. In the Advanced Structural Analysis subject, the theory of thin-walled beams is covered. Thin-walled section beams subjected to axial, bending, shear and torsional loads are analysed for the resulting displacements and direct and shear stresses. Therefore then, in verifying the finite element solutions to the considered case studies, the student is able to perform hand calculations based on the classical structural theories.

BASIC MODELLING TECHNIQUES

Before the students were allowed on the computer terminals, a three hour session was spent in the discussion and explanations relating to the basic essentials of the modelling techniques. The points listed below were meant to be of a general application, but the students found it helpful to have a structure to relate the problem to; in this context, the truss structure shown in Figure 1 was used as an example.

Axes System

In the first instance, the definition of the global axes system is required as the basic origin. Here, the three different basic axes system were explained: cartesian, cylindrical polar and spherical . In this case, the right-handed cartesian axes system was used to define the global axes system. However, it was pointed out to the students that in complex geometry, it may be more convenient and quicker to define parts of the geometry in different axes system. The point was made that it is possible for the user to define any axes system so long as one node on the user defined axes system refers back to the one of the three basic axes system. It is very easy for students to runaway with the impression that the cartesian axes system is the only available (in view of the nature of the displacements generally being given in cartesian axes system).

Element Related Properties

The specification of the element associated properties followed next. The calling of a particular element from a suite of element library had to be explained, for subsequent element and material properties depended on the choice and capability of the element. It was strongly emphasised that the correct choice of element is an important feature of the finite element modelling. It was next explained that the element local axes system is defined by the topology of the element. It was emphasised that this was an extremely important aspect of the element definition, as all related force and stress output are usually given in the local element axes. The section geometries required were looked into next. Here, the necessity of defining the appropriate section properties for the chosen element was explained. For example, in the truss structure of Figure 1, only the section area for each element was required for the chosen element; properties like the second moment of areas were not needed. Here again, the specification of appropriate section properties requires structural judgement on the part of the user, and therefore students were discouraged to make random choice of element without appreciating the corresponding fundamental behaviour of the element and its formulation.

Material Properties

An isotropic material only was considered for all the analyses. The definition of element material properties was illustrated by defining a "new" material with user defined properties, rather than using the default pre-programmed properties in the package. This was done intentionally to show the student the facility of defining an element material. The definitions of an orthotropic element material (as applicable to fibrous composite materials) was discussed in general without any detailed attention. The intention of mentioning orthotropic material was to dispel the misconception among the students that all materials are of an isotropic nature. In the case of linear static analyses, only two independent elastic constants were required: Young's modulus and the Poisson's ratio values. It was brought to the attention that the same material module would be used to define additional properties if other analyses are to be performed; properties like the damping coefficient for dynamic analyses, and coefficient of thermal expansion for thermal analyses.

Load Application

The methods of load applications in a finite element model was explained. The concept of applying loads at nodes, and the equivalent nodal loads to replace a uniformly distributed load was discussed. It was emphasised that generally speaking loads are defined on the model in the global axes, although it is possible to define the load direction in a user defined local axes system. Also explained was the idea of enforced displacements at nodes as being equivalent to a load application on the structure, and how this could be used in some cases to "zoom" in on a sub-structure model with a refined mesh for a more accurate analyses.

Boundary Conditions

Finally, the method of nodal restraints were discussed. This was discussed in two aspects. Firstly, the necessity of restraining the correct degrees of freedom appropriate to the element used. For example then, for a one degree of freedom line element, the only possible restraints are the two displacements (horizontal and vertical) at any node; there can no restraint on a rotation at a node as the element has no rotational degree of freedom. The second aspect was the absolute necessity of a minimum number of nodal restraints in the appropriate degrees of freedom to avoid rigid body motion of the structure for the static analyses.

CASE STUDIES

Case studies of simple structures were undertaken to illustrate the features of the finite element method. The structures considered were modelled by line, membrane and shell elements. No attempt was made to model a structure using a three-dimensional solid element. Also, a conscious decision was made not to use any automatic data generation or graphic facilities of the package. This was done so as to allow the students to get the feel of the method by manually modelling the structure, and manually scanning the output for inspection of results.

Tension Line elements

The simplest of the line element library was chosen, this being an element with one degree of freedom in the element axes, that is the axial displacement. The corresponding output is also one dimensional in terms of axial compression or axial tension force carried at the end nodes of each element. A statically determinate simple three member plane truss was modelled by three elements as shown in Figure 1. The geometry of the truss was chosen so as to keep the hand calculations to a minimum.

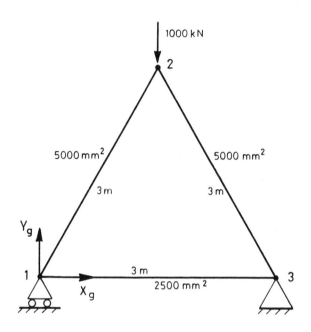

Figure 1. A three member plane truss structure modelled by tension line elements.

The inclination of the elements was deliberately chosen to highlight the appreciation of element and global stiffness considerations. The horizontal member 1,3 referred to in Figure 1, has its element axes coincident with the global structure axes system (Xg-Yg). The inclined members 1,2 and 2,3 (see Figure 1), would have different element and global axes system. The geometry and sectional areas of the three members are given in Figure 1; a constant Young's modulus value of 200 GPa was used. The member areas were also deliberately varied to demonstrate to the students the facility of using different element section properties in a model. This apparently simple example very competently and suitably illustrated the salient basic principles of finite element modelling, and helped the students put to practice the modelling techniques discussed earlier. The effect of the element topology in terms of the definition of the element axes and correct interpretation of the force/stress output was highlighted by this model. The students were asked to model the same structure by using a different topology for an element. For example, in member 1,3 referred to in Figure 1, the element topology was defined in the order of node 1 followed by node 3; this implied the element axes system (xe-ye) has the origin at node 1 with the positive element x-direction in the direction of node 1 to 3. This run gave an output of a negative axial force at node 1 and a positive axial force at node 3, thereby implying an axial tension force in member 1,3 (see Figure 2a).

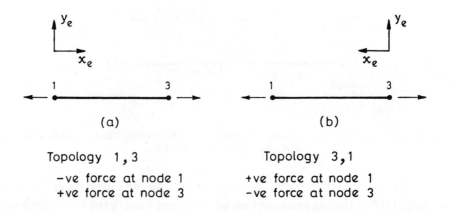

Figure 2. Importance of element topology definition and correct interpretation of forces.

The students were then told to rerun the programme with the element topology for member 1,3 now reversed, with node 3 now being the first and node 1 being the second in the element topology definition; this topology gave a positive force at node 1 and a negative force at node 3 (Figure 2b). This apparently is different from the first case, but this is consistent with an axial tension force in member 1,3 with the local axes element definition. This simple exercise helped to clearly illustrate the importance of the element topology definition.

Bending Line Elements

The next element considered was the six degrees of freedom line element. A simple statically determinate cantilever frame loaded out-of-plane as shown in Figure 3 was to be modelled. Both the members had the same sectional properties. The origin for the global axes system (Xg-Yg-Zg) was chosen to be at node 2.

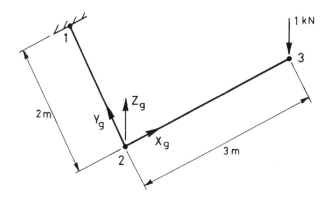

Figure 3. A two member cantilever frame modelled by bending line elements.

The two simple statically determinate structures of a truss (Figure 1) and a cantilever frame (Figure 3) were chosen to highlight the importance of the engineer's knowledge of the behaviour of structures under a given load system, and consequently then the judgement on the choice of element. For example, using the tension line elements on the cantilever frame would not be correct, as these elements would only carry axial forces with no other modes of loads allowed. Whereas the cantilever frame of Figure 3 will be subjected to a constant shear force in the Zg-direction, zero torque and a linearly varying bending moment about the Yg-axis in the member 2,3; a constant shear force in the Zg-

direction, a constant torque about the Yg-axis and a linearly
varying bending moment about the Xg-axis in the member 1,2.
The force and moment distributions in the members are shown
in Figure 4.

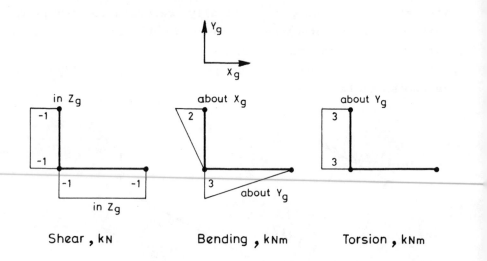

Shear , kN   Bending , kNm   Torsion , kNm

Figure 4. Forces and moments in members of cantilever frame
due to load system shown in Figure 3.

The students having appreciated the need for a line
element to have more than one degree of freedom, the six
degrees of freedom were then discussed: axial force, two
vertical shears, two bending moments and the torque. It was
brought to their attention that there are bending line
elements available which also take into account the
transverse shear effects on the deflection of members.   It
was pointed out that whilst it was incorrect to use the
tension line elements for the cantilever portal frame
structure, it is technically correct to use the bending line
elements on the truss structure, for in the bending line
elements the axial forces are allowed for in the analysis.
This was where an engineer's judgement was required to make
an efficient and economical choice of element. For using the
bending element for the truss structure would require
additional cross-sectional member data and computer run time,

whereas using the line tension line element would only require the cross-sectional area value for each member and less computer run time.

Membrane Elements

The correct modelling of the truss and cantilever portal frame structures gave the students the practice needed in the basic essentials of finite element modelling and using the available PAFEC package on the Apollo workstation systems. The students were then introduced to the two dimensional membrane elements. Emphasis was placed on the eight node quadrilateral elements. It was explained to them that the eight node quadrilateral elements were best for all round use, whereas the use of three and six node triangular elements should be kept to a minimum, or avoided if at all possible, because of their relatively poor performance. There was no time available to allow the students to prove for themselves the relative poor performance of the triangular elements.

A rectangular section cantilever beam under a tip load at the free end was the structure considered (Figure 5) for the membrane element modelling. A Young's modulus value of 200 GPa and Poisson's ratio of 0.3 was used.

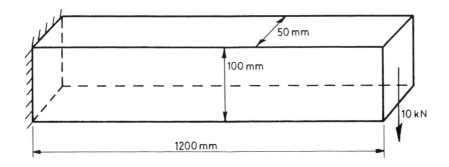

Figure 5. A rectangular section cantilever beam to be modelled by membrane elements.

The plane stress system of the structure through the width of the section was made clear to the students, thereby enabling the vertical plane of the structure to be modelled by membrane effects only, with the element thickness being the width of the section. The origin for the global axes system (Xg-Yg) was chosen to be at the bottom left corner of the beam. The most basic and crudest of the model was tried first to get the students going. This was the four node quadrilateral element shown in Figure 6, labelled as Mesh 1.

The concept of the equivalent nodal loads to simulate a distributed load was discussed here. No attempt was made to prove mathematically the equivalent nodal load expressions, but only the results quoted; with two corner nodes along the an element side, the load was divided in two equal parts, and for three nodes (two corner and a mid-side nodes) along an element side, the ratio of 1:4:1 is used. The students were encouraged to experiment for themselves for not using the correct nodal load ratios. Thus for example, in one run, the total load was applied at one node only (of Mesh 1, Figure 6), top or bottom node at the free end, and the vertical displacements for this run were compared with the results from a run in which nodal loads were correctly applied (that is, two equal load distribution at the nodes at the free end). The correctly applied nodal loads gave equal vertical displacements, whilst the incorrect nodal load application model gave unequal vertical displacements.

The importance of the element topology definition was particularly emphasised for the two dimensional elements as the element stress output is usually given in the element axes system. The element x-axis of all the elements were kept parallel (by correct ordering of the element topology) to the beam span direction. The direct stress due to bending would then be the element x-axis direct stress value.

The students were then required to model the same structure using eight node elements with different meshes, as shown in Figure 6. The aspect ratio of each element was kept the same within each mesh density, whereby the aspect ratio was defined as the ratio of the longest to the shortest side

of an element. Care was taken to ensure that the tip load was
applied uniformly along the free edge in the right ratios.

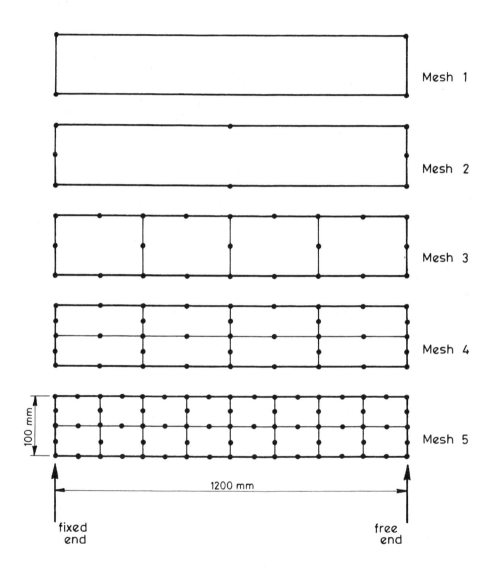

Figure 6. Different mesh densities for the rectangular
section cantilever beam problem.

For each run of different meshes shown in Figure 6, the
following results were to be noted: vertical displacements

along the nodes lying on the centre-line of the beam (neutral axis), except for Mesh 1 and Mesh 2 where the values were taken from the nodes on the top/bottom surface; direct stress values at the nodes on the top surface along the beam span; direct stress and shear stress values at the nodes through the section at the beam mid-span.

Shell Elements

The last type of elements considered was the two dimensional thin shell elements. The necessity of using shell elements as opposed to membrane elements was introduced to the students by asking them to model a thin-walled cantilever beam having a T-section profile, as shown in Figure 7.

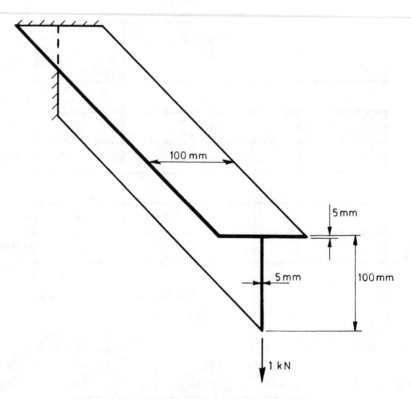

Figure 7. A thin-walled T-section cantilever beam to be modelled by membrane elements.

In the case of a rectangular section cantilever beam, it was possible to model the behaviour of the structure using membrane elements, but in the case of a T-section beam, a membrane load in one part of the section produces a bending effect on the other part in the section. Thus, for example, the shear in the web of the T-section is a membrane effect on the web; but this shear is now acting as an out-of-pane load on the flange at the flange/web junction of the T-section, thereby causing the flange to deform in a bending mode; and consequently then, the requirement for an element which caters for both the membrane and bending affects simultaneously on an element.

It was made clear to the students that the structure is still a plane stress problem, but membrane elements could not be used because of the three-dimensional nature of the geometry of the section. A crude mesh density as shown in Figure 8 was used. The necessity of ensuring nodal connectivity of the elements at the flange/web junction was discussed; this common junction line therefore necessitated at least two elements division across the flange. In addition to introducing the students to the shell elements, the other aim was to teach the correct interpretation of the stress output values at common nodes from elements lying in different planes; here the correct interpretation of the element axes system is of paramount importance as it is possible to define the local element x-axis in the flange, say, to lie along the beam longitudinal axis, whilst defining the flange element y-axis to line along the common beam longitudinal axis. In the case study considered here, students were asked to ensure that the element x-axis for both the web and flanges element were parallel to the common longitudinal beam axis direction. Vertical deflection at all the nodes at the free end were noted. Also noted were direct stress values along the beam longitudinal axis at all nodes at mid-span.

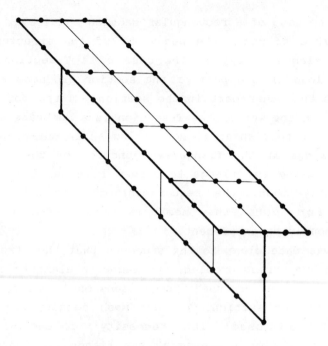

Figure 8. A mesh density for the T-section section cantilever beam problem.

DISCUSSION OF RESULTS

A selection of the results of the finite element runs for the case studies considered together with relevant discussion are presented here.

Tension Line Elements

The structure is a statically determinate one, and quick hand calculations were possible for a quick and correct comparison with the finite element values. Some students did get different values from their expected known solutions, and this brought up another facet of the finite element method. In that although a run has been "successful", it was emphasised that it did not mean that was what the engineer intended. For example, in cases of different answers between finite element and hand calculations, it was found that wrong properties were input, or the wrong restraints were used. When the correct solutions were obtained in terms of the axial forces, this gave confidence to the student as to the usage of the package.

The forces determined by hand calculation using the resolution of joints method (referred to as Theory in the table below) of the statically determinate structure of Figure 1, together with the finite element values of axial forces (referred to as Finite Element) are given in Table 1. A positive force implies tension, and a negative force signifies compression.

TABLE 1
Member forces (kN) in the plane truss structure.

Member	Theory	Finite Element
1,2	-577	-577
2,3	-577	-577
3,4	289	289

The correlation between the classical theory and the finite element method is extremely good as expected, because of the actual constant axial forces within the members, agreeing with the exact static element formulation based on the assumption of a constant axial force in a member. Such a good correlation between "theory" and finite element values were therefore explained to the student, least an impression be given that such exact answers are always possible.

Bending Line Elements
The correlation between the finite element values of forces, moments and torques in the two member cantilever frame of Figure 3 were found to be in excellent agreement with the values obtained from classical structural analysis which were presented in Figure 4. Again, the excellent correlation was

explained due to the exact static formulation of the bending line elements.

With this output, the students experienced difficulty in correctly interpreting the direction of forces, moments and torques consistent with the right hand element axes system. However, with the knowledge of the results from the structural analysis, most were able to work backwards to convince themselves of the correct output interpretation.

Membrane Elements

The students were asked to run the different meshes only after the inspection of results from the previous run. The aim was to get a satisfactory mesh for predicting three sets of output: vertical displacements along the beam span, direct stress along the beam span, direct and shear stresses through the beam thickness at mid-span. The hierarchy of accuracy required was in terms of displacements first, followed by direct stress, and finally the shear stress distributions.

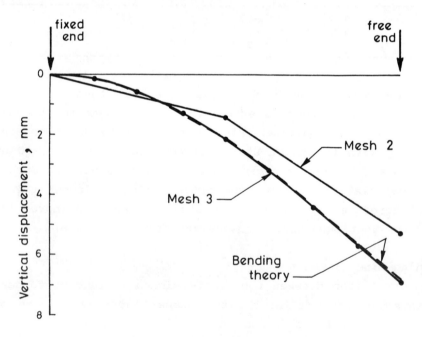

Figure 9. Comparison of displacements between bending theory and some finite element models.

As far as displacements were concerned, a refinement of Mesh 3 density was of sufficient accuracy as evident from Figure 9. The Mesh 1 model gave a tip deflection of 0.12 mm as compared to the predicted value of 6.9 mm. So as far as displacements were concerned, there was no need to refine the mesh any further. Here, the displacement results from different meshes showed the clearly the numerically approximate nature of the finite element method and its convergence aspects.

The students were then asked to look at the direct stress (on the top surface of the beam) variation along the span for the Mesh 3 model, since this model gave satisfactory displacement values. However, in inspecting the stress values at common nodes from adjacent elements, they were asked not only to plot the average of these two values, but also to indicate the variation between the stress values at common nodes from adjacent elements.

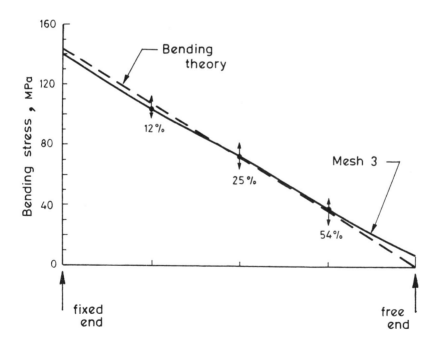

Figure 10. Comparison of direct stress between bending theory and some finite element models.

The variation between the stress values at common nodes from adjacent elementswas expressed as a percentage value, defined as the ratio of the difference between the higher and lower values to the lower value.

The average stress values variation along the span, shown in Figure 10, was in very good agreement with the bending theory. However, there is a large difference of values at common nodes from adjacent elements. Referring to Figure 10, a difference of 25% at mid-span and 54% at quarter-span from the free end. The relatively large figure of 54% difference at the quarter-span was attributed to the inadequate provision of elements for load diffusion effects, for there is only a single element between the load application nodes and the quarter-span position. A figure of a maximum of 10% variation was given to the students as being a satisfactory limit for the difference in stress values at common nodes.

So, in order to achieve a better accuracy of direct stress variation at common nodes, Mesh 4 and Mesh 5 were tried. Mesh 5 results were found to be satisfactory in that the results indicated a maximum of 11% difference between stress values at common nodes, with this maximum occurring at an element distance away from the loaded edge; again, this was attributed to the load diffusion problems. At other sections, all differences were within 5%, and this was regarded as being adequate. The students were also asked to check out the vertical deflections (although Mesh 3 did gave satisfactory results), and it was found that the refined meshes of Mesh 4 and Mesh 5 did give even better correlation with the predicted deflection to the third significant figure.

Here, it was pointed out to the students that the mesh refinement is dependent on the analysis sought. Thus, if only deflections were required, then the relatively crude mesh of Mesh 3 is satisfactory. Whereas, if direct stress values are required, then a refined mesh density (like Mesh 5) is needed.

Finally, the shear stress variation through the beam thickness was investigated. The results for Mesh 4 and Mesh 5 models are presented in Figure 11.

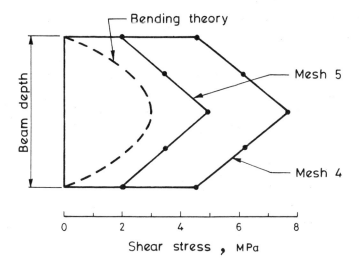

Figure 11. Comparison of shear stress between bending theory and some finite element models.

It is seen from Figure 11 that even the most refined of the models run (Mesh 5) does not gave a satisfactory representation of the shear stress variation. This comparison of displacement, direct stress and shear stress results for various models clearly illustrated the nature of the finite element method, in that the refinement used must be appropriate to the analyses required. The students were then asked to comment on the further mesh refinement required to achieve satisfactory shear stress results. By this time, most were able to correctly suggest that more element divisions would be needed across the beam thickness, although no further runs were made.

Shell Elements

The aim of modelling the thin-walled T-section beam was to introduce the students to the use of shell elements. By this time, they were confident of using the finite element package and have learnt the modelling techniques from the experience of the line and membrane element applications. The purpose here was therefore not to demonstrate any convergence of results for different meshes. Consequently then, only the basic mesh shown in Figure 8 was used.

For the mesh analysed, the tip deflection at the nodes at the free end from the finite element model was 12.6 mm, compared to the predicted theory, using thin-walled analysis, of 13.3 mm.

Also, the direct stress at the flange/web junction at mid-span were noted. Here, the of stress values from different planes at the same node required careful interpretation. For the point considered (flange/web junction at mid-span), there were six elements at the common node: two from the web and four from the flanges. The stress values from the web elements were 30.0 and 23.4 MPa, and from the flange elements were 27.7, 27.7, 20.9, 20.9 MPa; this gave an average value of 25.1 MPa (but with a maximum difference of 44% between the individual maximum and minimum values), whereas the predicted value using the bending theory is 25.0 MPa.

CONCLUSIONS

The practical approach of hand-on experience of modelling simple structures was popular with the students who felt they have learnt something useful, rather than having an all lectures based programme. The total time taken to cover the aspects discussed in this paper was thirty hours spread over ten three hour sessions. It is thought that the students spent on average an extra fifteen hours on the terminals by themselves without any staff supervision.

In terms of assessment, the step-by-step modelling of the rectangular section cantilever beam using different meshes was required as a report submission, in which graphical representation of the results and discussion was included.

AN FEM COURSE FROM CRADLE TO GRAVE

DAVID JOHNSON
Department of Civil and Structural Engineering,
Nottingham Polytechnic,
Burton Street, Nottingham NG1 4BU.

ABSTRACT

The progress, over a fifteen year period, of a final year undergraduate Engineering Analysis subject is followed from its initiation as a broadly-based, methodology-centred programme, through reorientation towards a structural bias, to its abandonment in favour of subjects with a design and construction orientation. The implicit changes in course aims and objectives responsible for the subject's creation, mutations and eventual cessation are considered and suggestions are made for explicit, focused and mutually agreed objectives in the interests of more logical course design.

Attention is also given to the aims which were specifically associated with Engineering Analysis. In this case, the subject is taken as an example of the effect that an evaluation of aims in relation to learners' attributes can produce in the content, operation and assessment of a subject which used the finite element method as a central linking feature. The importance of including behavioural objectives in subject aims is stressed and the Engineering Analysis example is used to indicate how such behavioural objectives were sought when defined as an improvement in the structural understanding of continuum elements and structural dynamics.

INTRODUCTION

Course Aims and Objectives

The requirements of validating bodies such as the Institutions of Civil and Structural Engineers' Joint Board of Moderators normally ensure that educational courses are endowed with a stated set of aims and objectives. The aims of the undergraduate course offered by my own department at Nottingham Polytechnic, for instance, are considered to be:

- to foster the intellectual, practical and personal development of the students by providing a good academic background together with sound practical experience. More specifically, the course aims to provide a sound basic education in the theory and practice of civil engineering and thus to prepare graduates for an interesting, challenging and useful career in society......

These aims are accompanied by a two-page list of objectives divided into Intellectual, Academic (subject-matter), Practical and Personal categories.

In practice, aims and objectives such as these are of little direct assistance in making *macro* decisions on course curriculum The aims tend to be so broad and the objectives so many that virtually any decision can be justified as meeting certainly the aims and probably a reasonable number of the objectives. Some aims and objectives, therefore, must be taken to be more equal than others. This pecking order, which varies with time and the personalities involved, will, together with practical constraints, determine the educational decisions which are taken.

Subject Aims

The syllabus for each subject area is also usually accompanied by *micro* aims which often amount to little more than a summary of the syllabus or are as self-evident as the aims of a Further Structural Analysis topic which were:

> *to extend previous studies to enable a student to be able to analyse more complex structures.*

Such aims tend to be too narrow and to focus on knowledge outcomes only to the exclusion of intellectual and behavioural benefits that may accrue from the learning process.

Aims and Objectives of this Paper

The purpose of the present paper is to reflect on the effects that trends in the emphasis given to educational aims and objectives produced on an the operation of a specific subject area in structural analysis which included, at its apogee, a substantial finite element component. The effect that changes of emphasis had on course objectives is related specifically to the creation and eventual demise of the subject. A detailed description is also given of the impact of a re-evaluation of the micro aims on the content, teaching/learning strategies and assessment modes associated with the subject.

'ENGINEERING ANALYSIS' AT THE POLYTECHNIC OF EAST LONDON

Conception And Birth

At the then Polytechnic of North East London (now Polytechnic of East London), the undergraduate course in civil engineering (long established under the auspices of the University of London) was transferred to the care of the Council of National Academic Awards (CNAA) in the early-1970's. In accordance with CNAA validation procedures, a submission document was prepared which incorporated a new curriculum to replace the external London University programme. The course was to offer honours and pass degrees, the first two academic years being common, but with separate final years, selection for which was determined by performance in the second year sessional examinations. The honours final year was characterised by a more 'in-depth' treatment and assessment of core subject areas; a more demanding project requirement; and, the hero of this tale, an extra subject.

The aim of this additional topic was to ensure that the honours stream was distinct from the ordinary degree and it was clearly necessary that the course be of a recognisably high academic standard. In addition, there was no reason to favour one civil engineering speciality over another so that breadth of coverage was indicated which could have been analysis, design, construction or context (related studies) based. Whilst the reasons for the decision are not documented, it is probable that the subject conceived, born, and eventually christened Engineering Analysis, was chosen due to emphasis given by the CNAA to academic respectability. Respectability, at the time, implied a mathematical/scientific, bias which, taken with the availability of an appropriate specialist, dictated the choice of an analysis based course. It was intended that breadth should be achieved by focusing on analytical techniques of, say, differential equation solution, which would then be exemplified by applications to structural, hydraulic and geotechnical problems.

Youth And Adolescence

As operated its early years, the Engineering Analysis course reflected the interests of its subject tutor and became structural biased, providing a introduction to continuum mechanics and structural dynamics with some material on thin-walled structures also. The finite difference method was treated in detail and the finite element method was considered in outline without specific applications.

The original intention of providing a general, methodology based course therefore floundered immediately since it did meet the practical constraint of matching course design with staff expertise and inclination. The course certainly succeeded in clearly distinguishing the honours and pass streams since few of the pass degree students would have relished Engineering Analysis, although whether this indicated a lower attainment in Civil Engineering in the wider sense is, of course, debatable.

On inheriting the course after some five years of operation, the author followed the pattern established by his predecessor for an initial couple of years before individuality asserted itself and some experience had been accumulated on the operation of the course. In reappraising the course, it was intended to retain the structural bias, since experience indicated that, within the confines of a 3 hour/week, one year course, a coherent treatment of a single subject area was possible, but extensions to other areas would have resulted in a rather shallow and possibly confusing treatment.

The particular aspect of the course which was felt to be least successful was elasticity solutions by use of stress functions, since the extensive algebraic manipulations involved tended to obscure and limit any insights into the physical behaviour of the problem under investigation. Given contemporary trends in analysis, extension of the finite element treatment was clearly indicated but the mode of its implementation in the course needed to be considered carefully in view of the fact that the majority of the students intended to follow professional design/construction careers and the probability of more than the occasional student having occasional resort to finite element analysis was remote.

With this background, it was decided to concentrate on the promotion of a basic understanding of the structural behaviour of continuum elements and of the principal features of vibration phenomena. Once this decision was made then many aspects of course content, delivery and assessment fell into place.

Maturity

Content: With the focus of the course firmly established on concepts of structural behaviour, the content of the course, whilst covering much the same topics, was restated to establish the primacy of structural action and the secondary importance which was to be given to analysis techniques. As restated, the course covered:

- Plane stress and strain elasticity theory using the finite element method to analyse stress concentration, deep beam and shear wall structures.

- Uniform torsion of non-circular solid sections and of open and closed thin-walled sections with reference to concrete core and box girder construction.

- Plate theory applied to bridge and floor slabs by a variety of numerical methods.

- Membrane theory of thin shells applied to axisymmetric retaining and roofing structures through closed-form solutions and bending theory applied to circular cylindrical tanks by the finite element method.

- Dynamic behaviour of two-dimensional line structures including the determination of natural frequencies for beams and frames.

- Computer modelling of two and three-dimensional line structures.

Finite element theory, including its stiffness method guise for line element structures, was involved directly or indirectly in each of the major topic areas and acted as a unifying factor. Nevertheless, finite elements were viewed as a purely structural tool, with the minimum of time spent on the theoretical basis of the method. Some finite difference applications (for plates and torsion) were retained. This was done partly for variety, partly to provide a direct application of governing differential equations and partly because the direct specification of stress boundary conditions required by the method was felt to focus attention on a physical aspect that could easily be overlooked using finite elements.

Delivery: The change to a behavioural focus was greatly assisted by the opportunity to produce a text-book [1], which covered the full content of the course apart from the computer modelling of line structures. Use of the book on the course enhanced both the quality and quantity of the formal contact time spent on the course. Much of the detailed theory was transferred to independent student study following an introduction which stressed the salient physical principles, practical applications and limitations of the theory. This was subsequently reinforced by a review session in which points of difficulty were discussed and extensions outlined. In this way the proportion of contact time devoted to formal lectures was reduced from around 50% to 33%.

The non-formal contact time had previously been devoted to problem solving sessions and a laboratory programme. The practicals comprised exercises on strain rosettes, Saint-Venant torsion, plate bending and membrane action in shells. Although it was of a traditional theoretical verification nature, this work was retained in full in order to enhance experimental skills and also to reinforce the theory and to demonstrate theoretical limitations. The range of loading for the plate, for example, was such as to just extend into the non-linear, membrane action range and the shell was clamped at its base and abruptly flattened at its apex so that bending effects were present at these extremities. Students were not forewarned of these features, but were asked to comment in detail on any discrepancies between theory and experiment. Structural dynamics was not included in the experimental programme but physical illustration was provided by the use of film and video [2]. Problem solving sessions were continued on a reduced scale but were accompanied by case study exercises, to which the saved formal contact time was also devoted.

The computer modelling of line structures was operated essentially solely on a case study basis in the manner which has been described previously [3]. A typical case study, based on a dome roof analysis described by Thomson et al. [4], is provided as Appendix 1. This was tackled as a piece of independent learning by students, supplemented only by some general comments on modelling along the precepts recently set out by Macloed [5]. A typical continuum finite element case study of a concrete cylinder splitting test is given as Appendix 2. Initially, a hand solution using a coarse, constant strain triangle mesh was tackled as a class exercise, groups of students dealing with a single element each and using a computer for the linear equation solution only. Although time consuming and on the tedious side, this hand solution was accompanied by exhortations to base equation formation on equilibrium considerations, to relate displacements to loading and restraints, and to compare the effectiveness of principal and component stresses in reviewing the physical reasonableness of the solution.

Following the hand solution, the same problem was solved using a commercial package so that the effects of element and mesh refinement could be studied and more detailed plots obtained on which a fuller description of the physical response could be based. For the first few years of operation of the subject, students performed the package analysis as an individual exercise, but subsequently this was done as a class demonstration to avoid package specific training. The time saved was devoted to a general consideration of modelling skills along the lines set out in the NAFEMS primer [6].

Assessment: Logically, the introduction of a case study element with the requirement of reports containing a substantial individual discussion element should have been accompanied by increased recognition of coursework in the assessment process. However, the straightjacket of the three hour terminal examination proved impossible to unlace and the best that could be achieved was to reduce the number of questions and provide partially computed results so that more realistic situations could be examined for which a qualitative appreciation of at least a part of the solution was expected. Typical questions for a line structure computer model and a continuum model, respectively, are included in Appendix 3.

Demise

Declining student numbers, combined with the decision not to accept honours degrees as an academic prerequisite for professional qualification, lead to the pass stream of the degree course becoming non-viable in the late 1970s. The two streams were therefore combined and a new combined course structure was prepared as part of CNAA review procedures. With its original rationale as an honours discriminant gone, it is not perhaps surprising that Engineering Analysis was under threat. However, without any amendment of the stated aims of the course, the fifteen years or so since its establishment had seen a distinct change of emphasis towards a design and construction orientation. Increasingly, also, a degree of student choice had been considered desirable, especially in the final year, and a topic seen as being academically demanding is unlikely to find favour in such a system. Thus implicit considerations sealed the fate of the Engineering Analysis, which was excluded from the revised degree structure.

CONCLUSIONS

If decisions on course structure are to be made rationally and feasibly, then statements of course aims and objectives need to be as concise and limited as possible and to focus on the essential character of what it is intended to achieve within the practical constraints which obtain. Ideally, these aims should be developed jointly by the teaching staff involved in the course so that all are aware of, and committed to, the objectives which are established. It was, for example, lack of appreciation of the broad role of the Engineering Analysis topic which lead to its diversion into a structurally biased subject.

Subject aims should also be brief but should concentrate on both the expected behavioural outcomes as well as the knowledge results which are desirable, taking into account the abilities, prior knowledge and aspirations of the learners. The restatement of the aims of the Engineering Analysis course in terms of behavioural objectives, which recognised the primarily professional aspirations of the learners and their need and preference for a physically rather than a mathematically based approach, was, for instance, extremely helpful in formulating the subject to meet these requirements.

REFERENCES

1. Johnson, D., *Advanced structural mechanics - an introduction to continuum mechanics and structural dynamics,* Blackwell Scientific Publications, Oxford, 1986.

2. Johnson, D., Watch it move!, *Structural Engineering Education Newsletter of the Institution of Structural Engineers,* 1991, No. 2, pp. 3-4.

3. Johnson, D., Why not *teach* computer analysis?, *The Structural Engineer,* 1989, **67**, No. 13, pp. 243-244.

4. Thomson, G., Schleyer, G.K. and Tooth, A.S., A design philosophy for large storage tank braced roofs, *The Structural Engineer,* 1987, **65B**, No. 3, pp 49-53.

5. Macleod, I.A., *Analytical modelling of structural systems,* Ellis Horwood, Chichester, 1990.

6. NAFEMS, *A finite element primer,* DTI, Glasgow, 1986.

APPENDIX 1 - DOME ANALYSIS CASE STUDY

Brief

The braced dome roof shown in Figure 1 has a roof radius of 29.28m and covers a 36.60m diameter tank. The roof comprises 40 meridional rafters, connected at the crown to a fabricated square hollow section centre-drum of radius 1.10m, four units of wind bracing, and five polygonal rings of bracing of radii 4.38, 7.64, 10.92, 13.65 and 15.55m. Initial design sizes for the rafter and the ring bracing are a 127x76x14kg/m joist and a 80x80x12kg/m angle respectively. The roof is subject to a live load of 1.2kN/m^2 of projected plan area in addition to the self-weight of the structural members and a 1.0kN/m^2 of roof area allowance for sheeting.

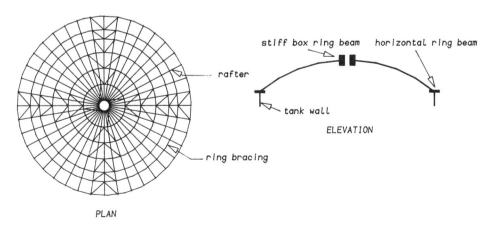

Figure 1. Dome plan and elevation.

Taking account of symmetry, a single rafter is to be subjected to a computer analysis using the model shown in Figure 2, in which the cross-sectional area of the equivalent ring bracing members, A_t, may be related to the actual cross-sectional area of the bracing, A_b by $A_t = \Theta A_b$ where Θ is the angular separation of adjacent rafters (radians). The tank wall may be presumed axially stiff but rotationally flexible.

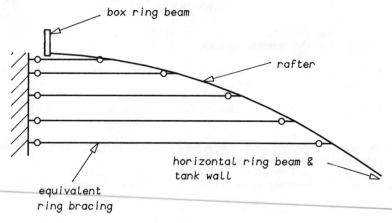

Figure 2. Analysis model.

Establish appropriate restraint and loading systems for the model and carry out computer analyses a. for the braced rafter and b. for the rafter alone without bracing. Provide plots of bending moment and axial load distributions for both the braced and the unbraced cases.

Discussion and Conclusions
Describe clearly the restraint and loading systems used, including any calculations. Explain the displacement pattern of the rafter. Compare the structural performance of the braced and unbraced rafters. Which system is structurally preferable, and why?

APPENDIX 2 - CONCRETE CYLINDER CASE STUDY

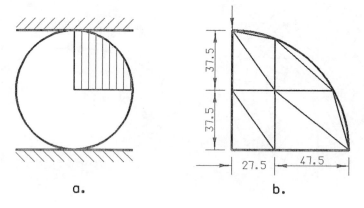

Figure 3. Concrete cylinder case study.

Brief

One form of concrete cylinder test is to subject a 150x300mm cylinder to a line load along a generator of the cylinder by placing the cylinder in a compressive testing machine as shown in Figure 3a. Taking account of symmetry and using the finite element net shown in Figure 3b, a quarter of the cylinder should be analysed for a total applied load of 300kN. (For concrete: E = 30kN/mm^2 , v = 0). A displaced shape diagram and a vector principal stress plot should be produced from the results obtained.

Discussions and Conclusions

Comment on the reasonableness of the calculated displacements and stresses. How do the central stresses compare with the closed-form solution: $f_x = 2P/\Pi D$; $f_y = -6P/\Pi D$; $f_{xy} = 0$, where D = cylinder diameter and P = load/unit length of cylinder. Under increasing load, how would you expect the cylinder to fail?

APPENDIX 3 - TYPICAL ASSESSMENT QUESTIONS

Frame Question

The frame shown in Figure 4a has been analysed by computer and some results from this analysis are given in Figure 4b. From the results, plot a bending moment diagram for the frame. State clearly the values of bending moment at each joint and, in addition, at the quarter-points of member 3-7.

If member 2-4 were omitted, so that the frame becomes three-pinned, determine the support reactions at joints 1 and 6 and hence plot a bending moment diagram for the modified frame.

State briefly the principal structural benefits produced by the inclusion of member 2-4.

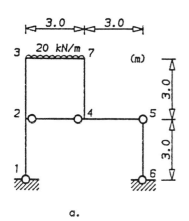

Member		F1	F2	M1	M2
		(kN)		(kN-m)	
2	3	-11.98	11.98	0.00	35.97
3	7	45.00	14.99	-35.97	-9.03
7	4	11.98	-11.98	9.03	-44.99
4	5	-14.99	14.99	44.99	0.00

b.

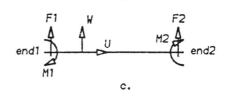

c.

a.

Figure 4. Frame question.

Elasticity Question

The tension strip shown in Figure 5a is bi-mettalic, its upper half being steel and its lower half being brass. The strip has been analysed by the finite element method and the displacement results are given, in part, in Figure 5b.

Calculate the stresses in elements 9 and 19 and relate these stresses to the expected behaviour of the strip. (t = 5mm; for steel: E = 202 000 N/mm^2 , v = 0.31; for brass: E = 102 000 N/mm^2 , v = 0.34).

80

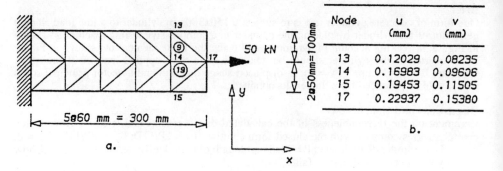

Figure 5. Elasticity question.

COMPUTATIONAL MECHANICS - CHAPTER 20 OR APPENDIX B ?

R D THOMSON
Mechanical Engineering Department
Glasgow University
Glasgow, Scotland

ABSTRACT

Since, in current professional practice, most engineering analyses are done computationally, a case is made for the introduction of computational methods and the finite element method in particular into the most elementary mechanics courses.   If this is at the expense of closed form solutions, then so be it!   Such an approach, it is argued, is necessary to ensure quality in the use of commercial finite element codes by practising engineers.   The proposal has been piloted in the Computer Integrated Engineering degree course at Glasgow University.   The response from students and employers has, so far, been encouraging.

NOTATION

ξ position within element in normalised coordinates

A area

E (Young's) elastic modulus

I second moment of area

k stiffness

L length in physical coordinates

N interpolation function

u displacement

x position in physical coordinates

Indexes in general

If a index is in parenthesis, it indicates a nodal value, eg:

$u_{(i)}$ represents the displacement at node i

$[a]^T$ the T indicates a transpose: if the matrix has a variance of one, the T indicates that it is a row matrix

Post-subscripts

a_i the i indicates a (covariant) component

$a_{(p)}$ the p is in parenthesis and so indicates the location at which a is measured

INTRODUCTION

Gilbert Strang, in the preface to one of the best textbooks (1) on applied maths to arise since computational mechanics came out of the closet, notes that "the teaching of applied maths needs a fresh approach." This is equally true of applied mechanics, in which most existing courses and texts tend to fall into one of three groups:

- the "elementary" group, whose subject matter is rather prosaic and whose mathematics is often outdated: not the stuff to fire the imagination of young Engineers

- the "advanced" group, in which the really interesting mechanics is discussed, often in modern terminology, but which assume that someone else has taught the elementary stuff

- the "computational methods" group, which assume an audience familiar with both of the above.

Since undergraduate engineering courses are very crowded and practising Engineers are short on spare time, much modern computational mechanics just never gets out to those who need it most (2). The very existence of NAFEMS is testimony to the problem. No engineering department can afford to regard computational methods and finite element analysis in particular as an appendix to the syllabus or, worse, a postgraduate topic.

The contention is then that computational methods should be fully integrated into a mechanical engineering curriculum from the outset. Numerical methods must be developed within an engineering rather than a mathematical context and applied as soon as possible, preferably when the engineering is relatively simple. This approach has been tested over the last few years on students studying for the degree in Computer Integrated Engineering (CIE) at Glasgow University, a course which is the role model for developments in other degrees offered by the Mechanical Engineering Department.

GENERAL OUTLINE

The Mechanical Engineering Department at Glasgow University offers degrees in Product Design Engineering (with Glasgow School of Art) and Mechanical Engineering as well as in CIE. The Department also services degree courses in Naval Architecture and Aerospace Engineering. All of these students are subjected to a common First Year course in Basic Engineering which includes Thermodynamics, Fluid Mechanics and Solid Mechanics. Beyond First Year, CIE students follow a separate syllabus from the rest, particularly in the Solid Mechanics with which this paper is concerned.

As in a "traditional" course, the primary aim of the mechanics course is to develop skills in the modelling of engineering systems, ie in the formulation of the governing equation(s) (GE) of the system. This requires that students develop a thorough understanding of the mechanics and physics of solids and considerable time is spent on emphasising such "traditional" ideas as free-body diagrams, equilibrium, compatibility, etc. In contrast, the solution of the GE is, at least to the Engineer, a purely routine exercise in mathematical manipulation. The formulation of the GE thus marks the point of departure from traditional courses with their emphasis on closed-form, graphical or manual tabular methods. The CIE course concentrates almost exclusively on computational solution methods and on finite element methods in particular. These are developed to solve even the simplest of system equations, even if these are amenable to closed-form solution, since an understanding of a generally applicable methodology is seen as more important than the generation of a particular solution. Indeed, many so-called "advanced" concepts are presented early in the course, not in the (vain) hope that students will immediately understand them or be able to reproduce the accompanying proofs but rather to admit the "spiral" approach to teaching in which the same ground is recovered in increasing depth several times in later years. This is seen as the only method which admits and allows the gestation period between initial presentation and final understanding.

Students are likewise given early exposure to industry standard software such as the Abaqus FE analyser, the Patran mesh generator and the Catia conceptual modeller. Although powerful packages capable of advanced applications, these are no more difficult for students to run than most so-called "teaching packages". They are also administratively convenient - having a single copy of a large commercial package mounted on a powerful multi-user platform reduces the housekeeping problem and allows policing of usage.

DETAILED STRUCTURE

In First Year, students attention is focussed exclusively on 1-D systems and the initial aim is to generate the governing equation for a uniaxial bar:

$$E \times A \times \frac{d^2 u}{dx^2} = 0$$

This is an ordinary differential equation (ODE) and presents none of the difficulty associated with the partial differential equations which model more general systems. The traditional approach in which this ODE is integrated in closed form to get the constitutive equation for the structure in question is thus rather sterile and does not readily extend to the more general case. In the CIE course, the aim is to develop general methodologies and so even this simple GE is recast into finite element form. The approach used is to define a displacement interpolation in terms of physical coordinates, viz:

$$u = u_{(1)} + \frac{x - x_{(1)}}{x_{(2)} - x_{(1)}} \times (u_{(1)} - u_{(2)}).$$

This is immediately transformed to the form:

$$u = [N_{(i)}]^T \times [u_{(i)}]$$

where

$$[N_{(i)}]^T = [\frac{1}{2}(1 - \xi) \qquad \frac{1}{2}(1 + \xi)].$$

The resulting approximate GE is then integrated using the method of weighted residuals (MWR) and Galerkin's method in particular. This leads to the stiffness matrix for the bar element as:

$$[k] = \frac{E \times A}{L} \times \begin{bmatrix} 1 & -1 \\ -1 & 1 \end{bmatrix}.$$

Just to reassure the doubters, the same result is obtained by elementary methods.

All of this might seem like using a sledgehammer to crack a nut but in wielding the sledgehammer, students are introduced to the concepts of shape functions, isoparametric elements, Jacobians, etc at a stage when the engineering is simple. Having thus encountered all of the buzzwords in First Year, students are presented in Second Year with an extension of the same general framework to cope with plane frames. Here the essential teaching point is the introduction of vectors to cope with the additional dimensions. These are denoted symbolically and in matrix format as appropriate but it is also useful to introduce at this stage both subscript notation and the summation convention and to apply these to such simple formulas as scalar products of vectors. Of course, in more-than-1-dimension, the equilibrium and compatibility requirements are more complicated and time must be taken out to explain such "traditional" matters.

Further into Second Year, students are given an introduction to beam theory which would not be out of place in a traditional course. Yet again however, a point of departure is reached once the moment-curvature relations are established for therafter students take a finite element approach to integrating the fourth order beam equation:

$$E \times I \times \frac{d^4 u_2}{dx_1^4} = 0.$$

The main teaching point here is the introduction of rotation as a degree of freedom in the matrix equations.

Galerkin's method would be clumsy for a fourth order equation and this motivates the introduction of the Principle of Virtual Displacements. It is fully admitted that a variational method based on virtual work would have afforded an algebraically simpler method than Galerkin's method for the formulation of the bar element stiffness matrix in First Year but virtual work is a sophisticated concept which is felt to be beyond the scope of a First Year course. Indeed, it is important that students appreciate that a variational principle is not always available and that recourse to MWR may be necessary.

Finally, in Third Year (there is no Fourth Year solid mechanics course in the CIE degree syllabus) the extension is made to 2-D continuum elements, the point at which many traditional courses are forced to abandon closed form solutions and adopt approximate methods. In such courses, students are faced with the double difficulty of developing a completely new methodology in the context of a relatively complicated engineering system. Far better to have laid the groundwork when the governing equations were themselves relatively simple and merely have to recap on concepts which are by now relatively familiar.

DISCUSSION

The approach adopted in the CIE course is open to, and has received, much criticism. A common complaint is that the "fundamentals" of mechanical science are lost by the emphasis on computational methods. But why should the use of approximate methods to solve the GE of a system be any less satisfactory to a pragmatic Engineer than the use of closed form solutions to mathematical models whose applicability to real problems is compromised by the simplifications needed to admit a closed-form solution?

It has also been argued that students will come to regard packages as black boxes and a substitute for understanding. But carefully structured tutorials in which students initially mimic the computational algorithm engender an appreciation of what the machine is doing. And in any case, the emphasis in Engineering must surely be on modelling the system and developing the governing equations, not on solving them!

In adopting this approach to mechanics, the traditional order in which topics appear has been radically altered. For example, since manual methods for plane trusses are not developed at all, trusses do not appear until vector transformations have been discussed. Conversely, tensors and subscript notation are introduced early and applied even to simple engineering systems. This pre-empts the misconceptions arising from the use of such terms as "stress vector" and avoids the need to develop new paradigms for more advanced applications, an approach which inevitably leads to intellectual indigestion. As a by-product, a self-consistent notation has been developed which is in close harmony with programming practice.

Mohr's circle, Macauley's method and other such techniques are simply dropped (at least as calculation methods).

CONCLUSION

The initial response from students and employers to this approach, which minimises the time by which graduates become profit-earning, has been positive. If the concern in NAFEMS about the quality of FE analyses can be directed at the graduates, the approach will have failed.

REFERENCES

1. Strang, G, Introduction to Applied Mathematics, Wellesley Cambridge Press, Wellesley, MA, USA, 1986.

2. Robinson, J, Finite Elements in Education, CME, March 1988.

PATCH TEST - A FUNDAMENTAL TOOL IN FINITE ELEMENT EDUCATION

ANDREW H C CHAN
Department of Civil Engineering
Glasgow University, Glasgow
UK G12 8LT

ABSTRACT

The primary objective of Finite Element Education is to equip the students with the ability of performing Finite Element Analysis for real physical problems. This can only be achieved with a proper understanding of the methodology. In this article, the Patch Test is established as a suitable and convenient tool for the teaching of virtually every important aspect of element behaviour except matters concerning material behaviour. Though only static linear elastic case is dealt with, most of the experience learnt can be applied to other situations.

INTRODUCTION

The primary objective of Finite Element Education is to equip the students with the ability of performing Finite Element Analysis for real physical problems. In order to perform such analyses, a practioner will immediately face the following questions:

Choice of interpolation function

Choice of material model

Choice of material parameters

The number of elements to be used

Type of boundary condition

Type of temporal discretization

Finite Element method, being an approximation to a mathematical theory that models the real physical reality, cannot deal with the questions of material model within its own framework. However, as shown in this article, the Patch Test introduced by Irons in 1965 [1] can help the users to acquire the necessary 'feel' for the decision of the other questions. In this paper, we shall concentrate on the static linear elastic case though criteria concerning stability, consistency and the rate of convergence have long been established for time stepping schemes [13] for dynamic condition.

In the following section, the history of Patch Test will be reviewed. Then the patch test as defined in Taylor et al in 1986 [9] is introduced. Discussions of the use of Patch Test in the understanding of the various aspects of Finite Element analysis are provided before coming to the conclusion that Patch Test is a fundamental tool in Finite Element Education. However due to limitation in space, few examples have been given in detail, the author would like to refer the enquirers to reference [9], from which most of the technical information of this article is derived, when such examples are needed.

History of Patch Test

Patch Test was originally introduced or invented by Irons. In the original conception, this test simply verified that an arbitrary 'patch' of assembled elements reproduced exactly the behaviour of an elastic solid material when subject to boundary displacements consistent with constant straining. This physical insight led him to develop a more formal test first presented in 1965 [1], which became a widely used procedure for checking of finite elements (and their coding).

Numerous publications on the theory and practice of the test ensued [2-5] and mathematical respectability was added to those by Strang [6]. The validity and the usefulness of the test was questioned in [7,8] and a shadow of doubt was cast. However, its applicability was upheld by Taylor et al [9] and in that publication, the patch test was generalised to contain the convergency and stability tests.

If correctly interpreted, the patch test is:
1. The necessary condition for finite element convergence.
2. A check on sufficiency of convergence.
3. The assessment of the asymptotic convergence rate of a particular finite element form.
4. A check on the robustness of the approximate algorithm.
5. A way to develop useful and accurate finite elements which may violate the continuity requirement.

In this publication, we are aiming at finite element education, application 5 of the patch test, though important to element developer, will not be discussed.

Formulation of the Patch Test

The patch test is essentially comparing the Finite Element solution with an exact analytical solution obtained from fundamental mathematical considerations. It provides a direct link from the mathematical theory to its Finite Element approximation. In order that the finite element approximation be useful, it must converge to the exact solution as defined by the mathematical theory under some asymtotic conditions. To ensure convergence, it is necessary that the approximation fulfil both consistency and stability conditions [10]. For a set of differential equations

$$A(u) \equiv L(u) + b = 0 \qquad (1)$$

where u is the independent variable e.g. displacement in structural mechanics, L is the differential operator acting on the independent variable e.g.

$$E \frac{d^2}{dx^2} \text{ in 1-dimensional bar}$$

and b is the forcing function on the system e.g. the gravity load on the bar. The differential eqn (1) is defined in the domain Ω together with the boundary condtions

$$B(u) = 0 \qquad (2)$$

For instance, for an 1-D elastic bar, the boundary condition can be fixed displacement

$$U = Uo \qquad (3a)$$

or traction condition

$$E \frac{du}{dx} = \sigma_o \qquad (3b)$$

at the boundary.

The finite element approximation is given in the form

$$u \simeq \hat{u} = \underline{N} \underline{a} \qquad (4)$$

where $\underline{N}$ are the shape functions defined in each element Ω_e and $\underline{a}$ are the unknown parameters.

By applying standard procedures [11] of finite element approximation the problem reduces in a linear case to a set of algebraic equations.

$$\underline{K} \underline{a} = \underline{f} \qquad (5)$$

The consistency requirement ensures that as the size of the element h tends to zero the approximation equation (5) and, where not enforced, its boundary conditions will represent the exact differential equation (1) and the boundary conditions (2) (at least in the weak sense).

Furthermore, since Finite Element Method is an approximation to a mathematical model of the problem, rate of convergence has to be known before the method can be applied successfully to practical conditions.

If the rate of convergence is low and the approximation usi:g a reasonable mesh is poor, it may require excessive effort e.g. mesh refinement in order to obtain a satisfactory result. The rate of convergence is established as follows:

When the element size h is sufficiently small, the error at any point becomes

$$|u - \hat{u}| = O(h^q) \leqslant C.h^q \tag{6}$$

where $q > 0$ and C is a positive constant, depending on the position. Both C and q are strongly dependent on the type of element used.

On the other hand, in order to obtain an answer from the numerical approximation, the approximation must be _stable_. The solution of the discrete equation system (5) must be unique and spurious mechanisms, which may pollute the solution, must be avoided.

In Taylor et al [9], 3 sets of patch tests have been developed to establish:

a. The consistency condition

b. The stability condition

and

c. The rate of convergence.

The Consistency Condition

Test A (Fig. 1a) is performed by simply inserting the exact value of the parameters $\underline{a}$ into the ith equation of (5) and verify that

$$\underline{K}_{ij} \; a_j - \underline{f}_i \equiv 0 \tag{7}$$

In Test B (Fig 1b) only the values of $\underline{a}$ corresponding to the boundaries of the 'patch' are inserted and a_i is found as

$$\underline{a}_i = \underline{K}_{ii}^{-1} (\underline{f}_i - \underline{K}_{ij} \; \underline{a}_j) \tag{8}$$

where j corresponds to the boundary nodes and i corresponds to the internal nodes.

In Test C (Fig. 2), the natural boundary condition ('traction' in case of structural application) is applied to the mesh with minimum boundary conditions. A solution is sought for the remaining $\underline{a}$ values and compared with the exact basic solution.

The stability condition

The Test C described in the preceding section can also be used to assess the stability of the patch. As only minimum boundary condition is imposed, any communicable mechanism (see e.g. [16]) can be revealed. A further single element test [14] (Fig. 2b) with minimum boundary condition can be used to reveal any

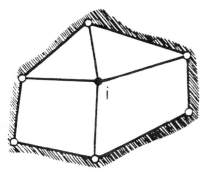 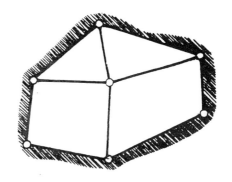

TEST A

$\underline{a}$ prescribed on all nodes

$\underline{K}_{ij}\underline{a}_j = \underline{f}_i$ verified at node i

TEST B

$\underline{a}$ prescribed at edges of patch

$a_i = K_{ii}^{-1}(f_i - K_{ij}a_j)(j \neq i)$ **solved**

FIGURE 1 PATCH TEST OF FORMS A AND B

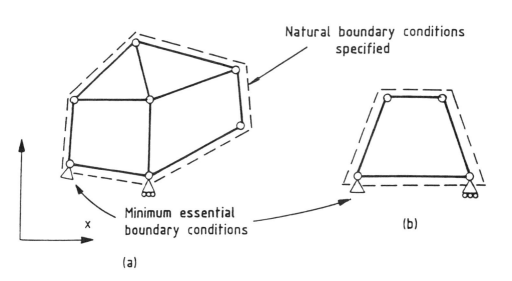

Natural boundary conditions
specified

Minimum essential
boundary conditions

x

(a)

(b)

FIGURE 2   a)   PATCH TEST OF FORM C
            b)   THE SINGLE ELEMENT TEST

singularity in the element.

The rate of convergence

While the patch tests discussed in the last two sections ensure (when satisfied) that convergence will occur, they did not test the order of this convergence. It is easy to determine the actual highest asymptotic rate of convergence of a given element by simply imposing in place of the exact basic solution an exact higher order polynominal solution. In addition, Test C in conjunction with a higher order patch test [5, 12] may be used to illustrate any tendency for "locking" to occur [11], e.g. Poisson ratio near $\frac{1}{2}$ for plane strain elasticity problems.

Patch Test as an Educational Tool

In this section, the use of patch test as an educational tool is discussed in detail. By considering various aspects of the finite element analysis, the way patch test can contribute to their understanding is illustrated.

Patch test is easy to perform

Patch Test can be performed by relatively few number of elements. This simplicity also applies to the irregular patch that is required to assess the performance of the element in a distorted condition. This can help to establish the confidence of the students towards performing Finite Element analyses. When higher order patch test is to be performed, a fine mesh may be required. Appropriate automatic mesh generation tool can be provided and its introduction is natural.

Patch Test is easy to mark

For the basic patch, the result is either satisfaction or not. For the higher order patch, the order of convergence can be assessed quite easily by examining the basic polynominal of the element. This removes the uncertainty of personal judgement on the accuracy of the solution. The lecturer involved can assign different patch test to different group of students and the results can be compared so that an overall picture can be obtained.

Patch Test provides the link between analytical solution and Finite Element Method

Students tend to think that what comes out of the computer must be correct. It is true up to a point. By using Patch Test, the students are confronted by two sets of answers. One is derived from the analytical solution which they can be asked to find out in simple cases and the other is obtained from the Finite Element approximation. It is straightforward to explain that the analytical solution is more accurate than the Finite Element solution within the appropriate mathematical framework. It is also the right moment to explain that the analytical solution is the solution of the mathematical description to the physical world and Finite Element is only an approximation to the mathematical approximation. It is hoped that through this process, 'blind faith' on the computer generated solution can be removed.

Patch Test and rate of convergence

Having removed the 'blind faith' in the finite element method, confidence must be rebuilt by demonstrating the convergence characteristic of the method. By analysing the same higher order patch test using a progressively finer mesh with increasing number of elements e.g. 1x1, 2x2, 4x4, 8x8 and so on, the rate of convergence of the element can be established. Usually the energy convergency rate is the easiest to establish and most accurately relect the average performance of the mesh (except in the case where singularity is found in the mesh). The rate of energy convergence can be approximated with the expression

$$\Delta E \simeq C_1 \, h^k \qquad\qquad (9)$$

where ΔE is the error (subtracting the approximated value from the exact value) in strain energy, C_1 is a positive constant, h is the element size and k is the rate of convergence. By plotting $\log\Delta E$ versus h, the value of k can be determined from the slope of the straight line drawn through the points. Usually the straight line relationship is valid only when h is sufficiently small.

In general, one can seldom establish the accuracy of a Finite Element solution to an actual problem with information obtained from only one mesh. We should always help our students to develop the habit of performing the analyses with more than one mesh where at least one of the meshes is much finer than the rest.

Patch Test and different mathematical theory used

If two plate elements are chosen to undergo the examination of the patch tests. One of them is based on Kirchoff thin plate theory and the other on the Mindlin plate theory. The student will notice that the thin plate will converge to the same non-dimensional answer no matter what l/t (length to thickness) ratio is used. However, the Mindlin plate element will give different answers for different l/t value. This would be puzzling at the beginning, but it can soon be revealed that Finite

Element does not model the physical world directly, it is only an approximation to the underlying mathematical theory. Since the two elements are based on different mathematical theory with slightly different basic assumptions, it is not difficult to appreciate that they should converge to a different answer. This will also help them to identify that Mindlin plate element is not necessarily more accurate although it may reach the thin plate analytical solution first.

Patch Test and order of polynomial

For a quadratic patch test (e.g. bending behaviour in 2-D elasticity), a rectangular 9-noded Lagrangian element can be shown to pass such a test with ease while it will take a lot of 4-noded bilinear elements with exact integration to come close to the exact answer. The student can learn from this patch test that 9-noded element will give better results than 4-noded element when bending (quadratic) behaviour is dominant. By studying the convergence rate of different elements within the same package, the students can form an objective judgement on the choice of elements in different situations.

Patch Test and distorted mesh

Although the 8-noded Serendipity element and the 9-noded Lagrangian element both have complete quadratic polynomial in the rectangular condition, only the 9-noded preserve the independence of the quadratic term in a linearly distorted condition (linear varying Jacobian matrix) (Fig. 3). This was first shown by Wachspress [17]. Both of these will pass the quadratic patch under rectangular condition, but only the 9-noded element will pass the quadratic patch test under a linearly distorted condition. The result of the two-element patch test under bending (loading 2 in Fig. 3) is given in Table 1. By performing patch tests for the conditions described above, the student can learn more objectively what a good mesh is.

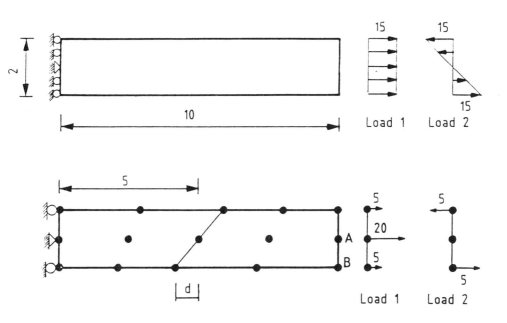

FIGURE 3   PATCH TEST FOR 8 AND 9 NODE ISOPARAMETRIC
QUADRILATERALS

TABLE 1

Quadratic Patch test for 8-noded and 9-noded plane stress elements
$(E = 100; \nu = 0.3)$

Element	Quadrature	d	v_A	u_B	v_B
8-noded	3x3	0	.750	.150	.75225
8-noded	2x2	0	.750	.150	.75225
9-noded	3x3	0	.750	.150	.75225
8-noded	3x3	1	.7448	.1490	.74572
8-noded	2x2	1	.750	.150	.75225
9-noded	3x3	1	.750	.150	.75225
8-noded	3x3	2	.6684	.1333	.66364
8-noded	2x2	2	.750	.150	.75225
9-noded	3x3	2	.750	.150	.75225
exact	–	–	.750	.150	.75225

Patch Test and loading condition

With loading applied in the ratio of 1/6, 4/6 and 1/6, a 9-noded element can pass a linear patch (i.e. simple tension condition) with ease. However if the loading is applied in the ratio of 1/3, 1/3 and 1/3, a refined mesh is required in order to achieve convergence to the analytical solution. The result of the two-element patch test under tensile stress (loading 1 in Fig. 3) is given in Table 2. This should drive home the importance of consistent loading in the finite element analysis.

TABLE 2
Linear Patch test for 9-noded plane stress elements
$(E = 100; \nu = 0.3; d = 0)$

Element	Quadrature	ratio	u_A	u_B
9-noded	3x3	1/6,4/6,1/6	.150	.150
9-noded	3x3	1/3,1/3,1/3	.144958	.160084
exact	–	–	.150	.150

Patch Test and 'Problem Elements'

Using the simple one element 'minimum constraint' test, the danger of 'reduced integration' [15] can be exposed for the 4-noded bilinear element with 1x1 Gauss Quadrature, 8-noded Serendipity element and 9-noded Lagrangian element with 2x2 Gauss Quadrature (see Fig. 4). It is well known that singular deformation mode exists for a single element. Though the mode for the 8-noded disappears when several elements are assembled, i.e. the singular mode is non-coimmunicable, the element should always be used with care. Also by assessing the rate of convergence for an exactly integrated 4-noded element with Poisson ration equals to 0.4999 [9,11] under plane strain assumption, the locking behaviour of this element at the incompressible limit can be revealed (Fig. 5 and 6). By introducing a few 'problem elements' in the coursework, the alertness of the students towards the use of finite element method can be raised.

Patch Test and Finite Element Packages

After graduating, students may be confronted with Finite Element Packages and elements which are totally alien to them. By performing the familiar patch tests which they have learnt in their course, a feel can be developed for the relevant elements in the package. Valuable information about the element characteristic e.g. reliability, stability, robustness, accuracy and rate of convergence can be established systematically.

Conclusion

When we teach our students the Finite Element method, it is our intention that they can handle their future analyses with confidence. We would certainly like to impart to them useful knowledge on, for instance, optimal sampling point for stresses and various non-linear applications. However, it is important to establish their basic understanding of the method before such advanced techniques are introduced. In this article, patch test has been established as a convenient and effective tool in achieving the understanding and feel for the finite element in order to gain such confidence. This kind of understanding and confidence cannot be easily established using few and complicated 'practical' examples. Thus the use of Patch Test (or simple comparisons between analytical solutions and finite element results) is fundamental to Finite Element education. It is sincerely hoped that Patch Test will find its way more often into elementary Finite element courses currently on offer.

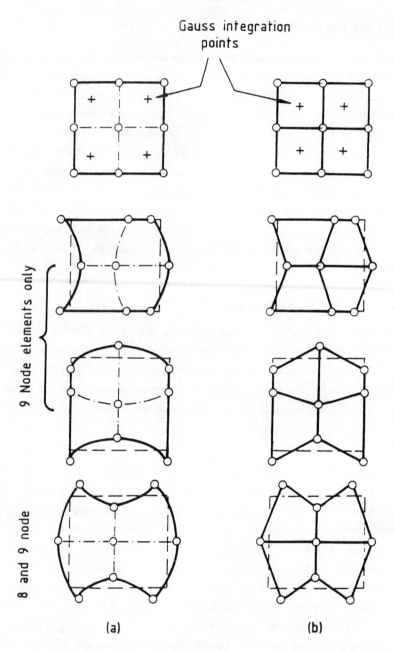

FIGURE 4   ZERO ENERGY (SINGULAR) MODES FOR 8 AND 9 NODE
QUADRATIC ELEMENTS (a) AND FOR A PATCH OF
BILINEAR ELEMENTS WITH SINGLE INTEGRATION POINTS (b)

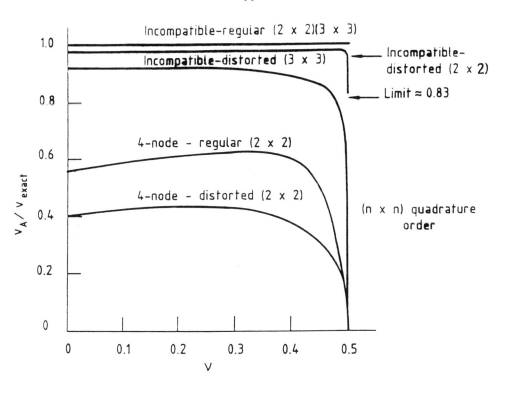

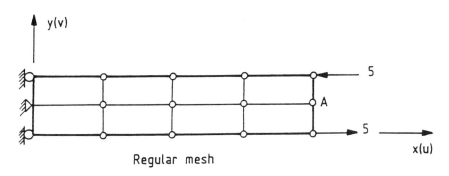

Regular mesh

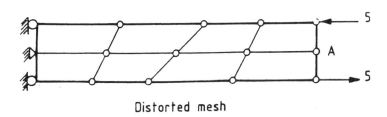

Distorted mesh

FIGURE 5   PLANE STRAIN 4-NODE QUADRILATERALS WITH AND
WITHOUT INCOMPATIBLE MODES (HIGHER ORDER PATCH
TEST FOR PERFORMANCE EVALUATION)

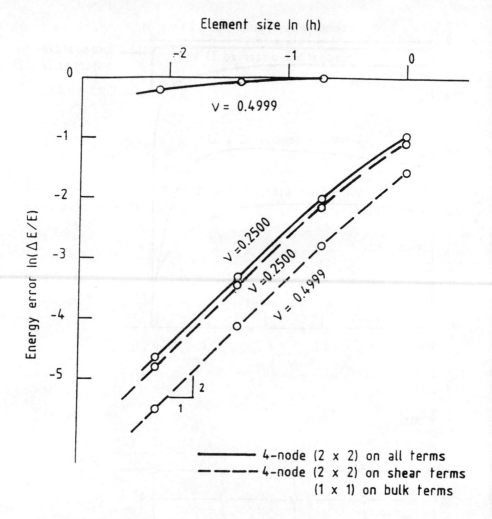

FIGURE 6   HIGHER ORDER PATCH TEST ON ELEMENT ROBUSTNESS
(SEE FIGURE 5) (CONVERGENCE TEST UNDER SUBDIVISION
OF ELEMENTS)

References

1. Bazeley G.P., Cheung Y.K., Irons B.M. and Zienkiewicz O.C., Triangular elements in plate bending – conforming and non-conforming solutions, Proc. 1st Conf. on Matrix Methods in Structural Mechanics, Wright Patterson AFB, Ohio, 1965, 547–576.

2. Irons B.M. and Razzaque A., Experience with the path test for convergence of finite element methods, Math. Foundations of the Finite Element Method, ed. A.K.Aziz, Academic Press, 1972, 557–587.

3. Fraeijs de Veubeke B., Variational Principles and the Patch Test, Int. J. Num. Meth. Engrg., 8, 1974, 783–801.

4. Sander G. and Beckers P., The influence of the choice of connectors in the finite element method, Int. J. Num. Meth. Engrg., 11, 1977, 1491–1505.

5. de Arantes e Oliveira E.R., The patch test and the general convergence criteria of the finite element method, Int. J. Solids Struct., 13, 1977, 159–178.

6. Strang G., Variational crimes and the finite element method, Proc. Foundations of the Finite Element Method, ed. A.K.Aziz, Academic Press, 1972, 689–710.

7. Stummel F., The limitations of the patch test, Int. J. Num. Meth. Engrg., 15, 1980, 177–188.

8. Robinson J., The Patch Test – is it or isn't it?, Finite Element News, 1, 30–34, 1982 – a collection of letters to the Editor by B.M. Irons, G. Strang and F. Stummel.

9. Taylor R.L., Zienkiewicz O.C., Simo J.C. and Chan A.H.C., The Patch Test – a condition for assessing FEM convergence, Int. J. Num. Meth. Engrg., 22, 39–62, 1986.

10. Ralston A., A First Course in Numerical Analysis, McGraw-Hill New York, 1965.

11. Zienkiewicz O.C., The Finite Element Method, 3rd Edn, McGraw-Hill, London, 1977.

12. Strang G. and Fix G.J., An Analysis of the Finite Element Method, Prentice-Hall, Eaglewood Cliffs, N.J., 1973.

13. Lambert J.D., Computational Methods in Ordinary Differential Equations, John Wiley and Sons Ltd., 1973.

14. Bergan P.C. and Hanssen, A new approach for deriving 'good' element stiffness matrices', Math. of Finite Elements and Applications, ed. J.R. Whiteman, Academic Press, 1976, 483–497.

15. Zienkiewicz O.C., Taylor R.L. and Too J.M., Reduced integration techniques in general analysis of plates and shells, Int. J. Num. Meth. Engrg., 3, 1971, 275–290.

16. Pugh E.D.L., Hinton E. and Zienkiewicz O.C., A study of quadrilateral plate bending elements with reduced integration, Int. J. Num. Meth. Engrg., 12, 1059–1079, 1978.

17. Wachspress E.L., Higher-order curved finite elements, Int. J. Num. Meth. Engrg., 17, 1981, 735–745.

ELEMEMTS WITHOUT WORK

EDWARD A.W. MAUNDER
School of Engineering, University of Exeter
North Park Road, Exeter, Devon EX4 4QF UK

ABSTRACT

A new formulation of the finite element method of analysis of elastostatic problems with displacement elements is described. This formulation has been developed for a final year undergraduate course for engineering students. This alternative to the more usual appeal to energy functionals, virtual work, or Galerkin weighted residuals aims to simplify an understanding of the mechanics of the method without loss of rigour. The approach uses duality concepts with equal weight being given to two roles of the shape functions in representing displacements and forces. To illustrate the process, stiffness coefficients are derived for 3 noded, 4 noded, and 8 noded membrane elements. A complete weighted residual equation emerges after the formulation which accounts for all types of force residual.

INTRODUCTION

When presenting the concepts of finite element methods for the first time to engineers (including students of engineering), it is considered that the following aspects should be addressed:

- physical significance should be given to quantities, functions, and arguments required to formulate the methods. This is necessary in order to develop an engineering feel for the method and its results, and in order to make judgements about the quality of approximate solutions (1).

- confidence should be maintained in the formulation of a model. Where approximations are necessarily made they must be clearly stated and understood, and not be hidden within principles which on close examination are seen to be used in an invalid way.

- the presence of errors in the response of a model should be recognized.

The simplest indicators of error are the residuals e.g. residual loads, and they should appear in the calculations. Furthermore residuals can be used in computing estimates of error energy norms, so they can have a greater role than merely to act as error indicators (5).

- the behaviour of a single element should be thoroughly understood before attempting to design or evaluate a multi-element model (4). To be in sympathy with this perceived need it is necessary to consider the complete response of an element e.g. distributions of both load and stress corresponding to node displacements of a displacement element. The usual approaches to formulating a finite element model rely on concepts of :
- energy,
- virtual work,
- weighted residuals.

However, when these concepts are considered with respect to the above aspects, the following comments can be made :

- energy : the principle of minimum potential energy requires the minimising of a functional - it produces the appropriate equations but tends to act as a 'black box'. Numbers go in and come out, but what does a total potential really mean? (2)

- virtual work : this principle is used to equate internal work to the external work of the nodal forces. This appears to involve an invalid use of the principle since nodal forces cannot possibly be a equilibrium with element stress distributions which contain only finite stresses. The step of transforming element stresses to distributions of element loads, and then evaluating discrete node forces is usually omitted. The student may not be aware that node forces form an approximation to an element response.

- weighted residuals : this usually involves the Galerkin method and is usually presented as a 'top down' formulation i.e. the complete model is considered first before the individual elements. Athough superficially this method appears to have a simple and appealing mathematical basis, it becomes unclear as to just which residuals are being accounted for (6), and physical appeal is lacking. Element stiffness matrices appear by analogy rather than by direct consequences of their behaviours (3).

This paper continues with an alternative formulation based on duality concepts already expounded for the heat conduction problem (7), and generally considered in relation to elastostatics (8).

DUALITY OF DISPLACEMENT AND FORCE FIELDS FOR A DISPLACEMENT ELEMENT

Displacements are restricted within an element to those fields defined by interpolating between nodal values. Restricting attention to 2-D membrane elements, one scalar interpolation (or shape) function N_i^e is defined per node i with the properties :

$$N_i^e = 1 \text{ at node i} \tag{1}$$
$$N_i^e = 0 \text{ at all other nodes}$$

This set of properties allows interpolation when only one of the nodes moves – say by vector displacement Δ_i^e. Then any point P of the element moves parallel to the node by :

$$\delta_p = N_i^e \, \Delta_i^e \tag{2}$$

when N_i^e is evaluated at P. Refer to Figure 1.

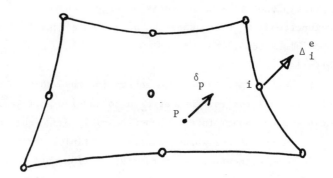

Figure 1 Displacement in a 9-noded element

In general when all nodes move, P is displaced by the linear combination of node vectors , $\delta_p = \Sigma N_i^e \, \Delta_i^e$, or in matrix form:

$$\{\delta_p\} = [N^e] \{\Delta^e\} \tag{3}$$

$$2\text{x}1 \quad 2\text{x}2n \quad 2n\text{x}1$$

where $\{\Delta^e\}^T = [\, \Delta_1^T \Delta_2^T .. \Delta_n^T]$ for an n-noded element.

An essential requirement for the displacement fields so interpolated is that no 'spurious' strains are developed in the element when representing rigid body movements.

Thus when nodal displacements conform to rigid body movements (zero strain), any other point P must likewise conform. The necessary and sufficient conditions on the set of element shape functions to represent all rigid body movements are :

$$\sum_i N_i^e = 1 \quad \text{at P} \tag{4}$$

$$\sum_i N_i^e \, r_i = r_p \quad \text{at P, where } r_p \text{ and } r_i \text{ are position}$$
$$\text{vectors of P and node } i \text{ respectively.}$$

A further consequence of these two conditions on the element shape functions is that they can also represent exactly any linear scalar function by interpolating from true nodal values. A bonus follows in representing displacement fields: all linear vector fields of displacement can also be modelled by the element. · In other words all constant strain states are included in the element's modes of behaviour.

Element displacements are represented by nodal values. The nodes indeed represent the connecting points (or terminals) to the outside world, whether it be to other elements in a mesh, or to applied loading.

The entities of displacement and force are dual to each other, and dual transformations can be used to represent displacement and force at any point P in an element. Thus a force q_p applied at P can be represented by a set of parallel forces applied to each node (this can be thought of as 'load dispersion') :

$$F_i^e = N_i^e \, q_p \tag{5}$$

where F is the force at node i. Refer to Figure 2.

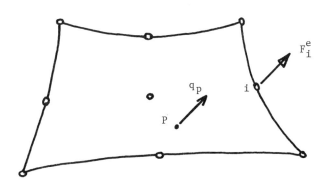

Figure 2 One of the node forces representing q_p

In matrix form :

$$\{F^e\} = [N^e]^T \{q_p\}$$

(6)

$$2n\times1 \quad 2n\times2 \quad 2\times1$$

$$\text{where } \{F^e\}^T = [F_1^T \ F_2^T \ \ldots F_n^T]$$

Thus provided force and displacement components are based on the same directions for axes, transformations between displacements and force are then contragredient (as is commonly the case in mechanics). In finite element terminology $\{F^e\}$ is consistent with $\{q_p\}$.

Now the properties of the shape functions which ensured rigid body displacements also ensure that the consistent forces $\{F^e\}$ are statically equivalent to q_p. This follows by considering :

$$\text{the resultant force} = \sum_i F_i^e = \sum_i N_i^e q_p = (\sum_i N_i^e)q_p = q_p \text{ when } \sum_i N_i^e = 1,$$

$$\text{and the resultant moment} = \sum_i r_i \times F_i^e = \sum_i r_i \times (N_i^e q_p) = (\sum_i N_i^e r_i)\times q_p = r_p \times q_p.$$

Furthermore the node forces do the same amount of work as q_p for an arbitrary displacement field defined by nodal displacements $\{\Delta^e\}$, since :

$$\sum_i (F_i^e \cdot \Delta_i^e) = \{F^e\}^T\{\Delta^e\} = \{q_p\}^T[N^e]\{\Delta^e\} = \{q_p\}^T\{\delta_p\} = q_p \cdot \delta_p$$

(7)

Consistent node forces are thus 'work equivalent' to an applied force q_p.

Distributions of applied forces such as body forces f^e and edge tractions T^e are represented by nodal forces by considering q_p as an infinitesimal limit of $f^e \delta v$ or $T^e \delta a$ and integrating over or around the element. Refer to Figure 3 and equation (8) following.

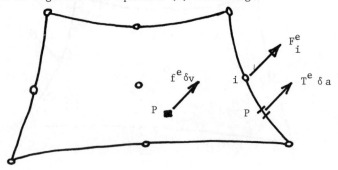

Figure 3 One of the node forces representing load distributions f^e or T^e.

$$\{F^e\} = \int [N^e]^T f^e dv + \int [N^e]^T T^e da \qquad (8)$$

The property of statical equivalence ensures that no force is lost or created by the dispersion process, and it should appeal to engineering 'intuition' - although load distributions are perhaps more often represented by a single resultant force or couple rather than a set of forces.

ELEMENT STIFFNESS COEFFICIENTS

The stiffness coefficients of interest here relate the nodal variables of displacement and force - both types of variable represent different aspects of the element behaviour.

Starting with nodal displacements $\{\Delta^e\}$, a sequence of action and response entities are now derived as follows :

$$\{\delta\} \quad = \lfloor N^e \rfloor \{\Delta^e\} \quad \text{interpolates element displacements.} \qquad (9)$$

$$\{\bar\varepsilon\} \quad = [B^e]\{\Delta^e\} \quad \text{derives element strains.} \qquad (10)$$

$$\text{where } [B^e] = \begin{bmatrix} \dfrac{\partial}{\partial x} & 0 \\ 0 & \dfrac{\partial}{\partial y} \\ \dfrac{\partial}{\partial y} & \dfrac{\partial}{\partial x} \end{bmatrix} [N^e] = [\partial][N^e]$$

$$\{\bar\sigma\} \quad = [H^e]\{\bar\varepsilon\} \quad \text{derives element stresses from the material stress-strain relations.} \qquad (11)$$

$$\{\bar f^e\} \quad = -[\partial]^T\{\bar\sigma\} \quad \text{derives body forces in equilibrium with stresses.} \qquad (12)$$

$$\{\bar T^e\} \quad = [C]\{\bar\sigma\} \quad \text{derives edge tractions in equilibrium with stresses.} \qquad (13)$$

$$\text{where } [C] = \begin{bmatrix} n_x & 0 & n_y \\ 0 & n_y & n_x \end{bmatrix} \text{ and } n_x, n_y \text{ are direction cosines of the edge normal.}$$

Note that all vector components in the above use common reference axes (X,Y).

Finally the force distributions $\bar f^e$ and $\bar T^e$ are represented by nodal forces :

$$\{\bar F^e\} = \int [N^e]^T\{\bar f^e\}dv + \int [N^e]^T\{\bar T^e\}da \qquad (14)$$

Now by starting with a unit displacement of one node the response $\{\bar F^e\}$ is a vector of node forces which are regarded as stiffness coefficients.

The vectors corresponding to all such unit displacements assemble to form the columns of the element stiffness matrix $[k^e]$, so that generally

$$[k^e]\{\Delta^e\} = \{\bar{F}^e\} \qquad (15)$$

This derivation of a column of the stiffness matrix is illustrated for three commonly used elements, the three node triangle, the four node quadrilateral, and the eight node quadrilateral. These examples show clearly the source of the conributions to each stiffness coefficient, and the influence of Poisson's ratio is also readily apparent. Refer to Figures 4 to 8.

However although this procedure makes clear the behaviour of an element, and is suitable for hand calculation, an alternative formulation of $[k^e]$ is normally used for computer programs. The relation between the formulations makes use of the element identity :

$$\int [N^e]^T\{f^e\}\,dv \;+\; \int [N^e]^T\{T^e\}\,da \;=\; \int [B^e]^T\{\sigma^e\}\,dv \qquad (16)$$

where a continuous stress field $\{\sigma^e\}$ is in equilibrium with continuous load distributions $\{f^e\}$ and $\{T^e\}$. The identity can be seen as a simple consequence of the divergence theorem applied to a stress vector weighted by a shape function :

$$\text{e.g} \quad \text{divergence } N_i^e \begin{Bmatrix} \sigma_x \\ \tau \end{Bmatrix} = N_i^e \left(\frac{\partial \sigma}{\partial x}x + \frac{\partial \tau}{\partial y}\right) + \left(\frac{\partial N_i^e}{\partial x}\cdot\sigma_x + \frac{\partial N_i^e}{\partial y}\cdot \tau\right) \qquad (17)$$

$$= -N_i^e f_x^e + \left[\frac{\partial N_i^e}{\partial x} \quad 0 \quad \frac{\partial N_i^e}{\partial y}\right]\{\sigma^e\}$$

From the divergence theorem

$$\int \text{div}\left(N_i^e \begin{Bmatrix} \sigma_x \\ \tau \end{Bmatrix}\right)dv \;=\; \int N_i^e (\sigma_x n_x + \tau n_y)\,da = \int N_i^e T_x\,da \qquad (18)$$

Hence

$$\int N_i^e f_x\,dv + \int N_i^e T_x\,da \;=\; \int \left[\frac{\partial N_i^e}{\partial x} \quad 0 \quad \frac{\partial N_i^e}{\partial y}\right]\{\sigma\}\,dv \qquad (19)$$

or dispersion of continuous loads in a particular direction to a particular node is also given by the integral on the right hand side. $\{f_x\}$ and $\{T_x\}$ are in equilibrium with $\{\sigma\}$. By similarly considering the orthogonal direction, and all the node shape functions, the complete matrix identity is obtained. The stiffness matrix now comes from:

$$\{\bar{F}^e\} = \int [B^e]^T\{\bar{\sigma}\}\,dv$$

$$= \left[\int [B^e]^T[H^e][B^e]\,dv \right]\{\Delta^e\} = [k^e]\{\Delta^e\}$$

Then
$$[k^e] = \int [B^e]^T[H^e][B^e]\,dv \qquad (20)$$

$\{\Delta^e\}^T = [1 \ 0 \ 0 \ \ldots \ 0]$

$\{\delta^e\}^T = \dfrac{1}{h} \ [y \quad 0]$

$\{\bar{\varepsilon}^e\}^T = \dfrac{1}{h} \ [0 \quad 0 \quad 1]$

Plane stress:-

$\{\bar{\sigma}^e\}^T = \dfrac{E}{h(1-\nu^2)} \ [0 \quad 0 \quad \dfrac{(1-\nu)}{2}]$

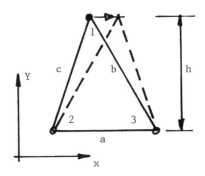

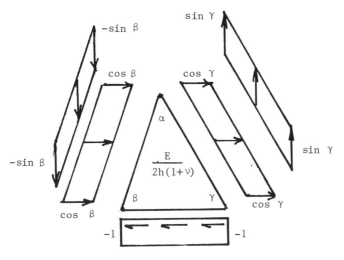

<u>Boundary tractions</u>

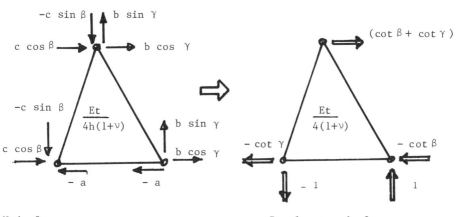

<u>Node forces</u>   <u>Resultant node forces</u>

Figure 4 3 noded triangular element: stiffness coefficients.

$\{\Delta^e\}^T = [1 \quad 0 \quad 0 \ ...0]$

$\{\delta^e\}^T = \frac{1}{4} \ [(1+x)(1+y) \quad 0]$

$\{\Sigma^e\}^T = \frac{1}{4} \ [(1+y) \quad 0 \quad (1+x)]$

Plane stress:-

$\{\overline{\sigma}^e\}^T = \frac{E}{4(1-\nu^2)} \ [(1+y) \quad \nu(1+y) \quad \frac{(1-\nu)}{2}(1+x)]$

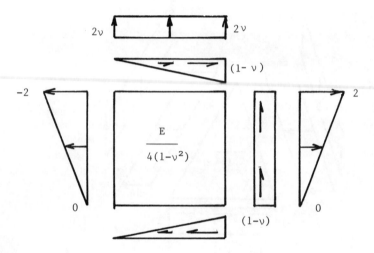

Boundary tractions

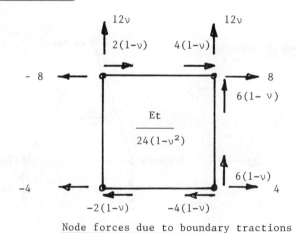

Node forces due to boundary tractions

Figure 5 4 noded square element: displacement of node 1

$$\{\bar{f}^e\}^T = \frac{-E}{4(1-\nu^2)} \quad [0 \quad \frac{(1+\nu)}{2}]$$

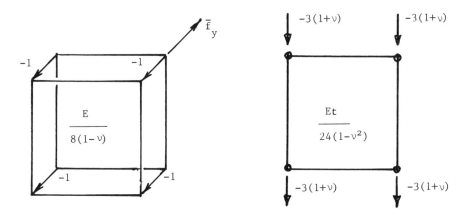

Body forces and equivalent node forces

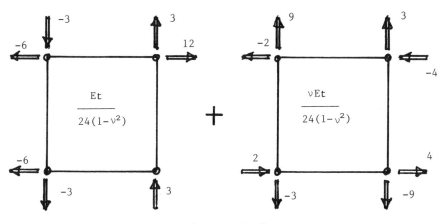

Resultant node forces

Figure 6 4 noded square element: stiffness coefficients

$\{\Delta^e\}^T = [1\ 0\ 0\ ...0]$

$\{\delta^e\}^T = \dfrac{1}{4}\ [(1+x)(1+y)(x+y-1)\ 0]$

$\{\bar{\varepsilon}^e\}^T = \dfrac{1}{4}\ [(1+y)(2x+y)\quad 0\quad (1+x)(x+2y)]$

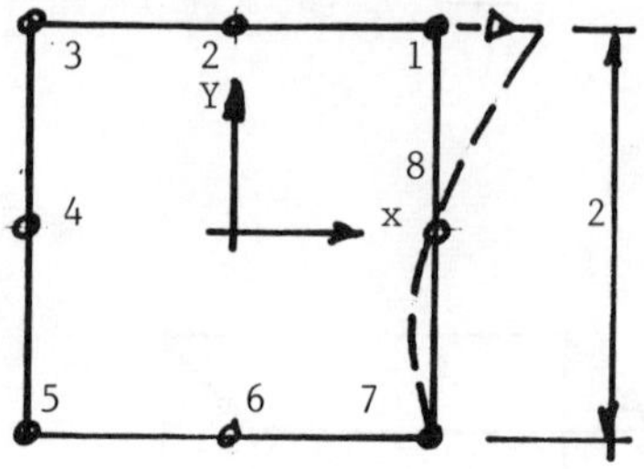

Plane stress:-

$\{\bar{\sigma}^e\}^T = \dfrac{E}{4(1-\nu^2)}\ [(1+y)(2x+y)\quad \nu(1+y)(2x+y)\quad \dfrac{(1-\nu)}{2}(1+x)(x+2y)]$

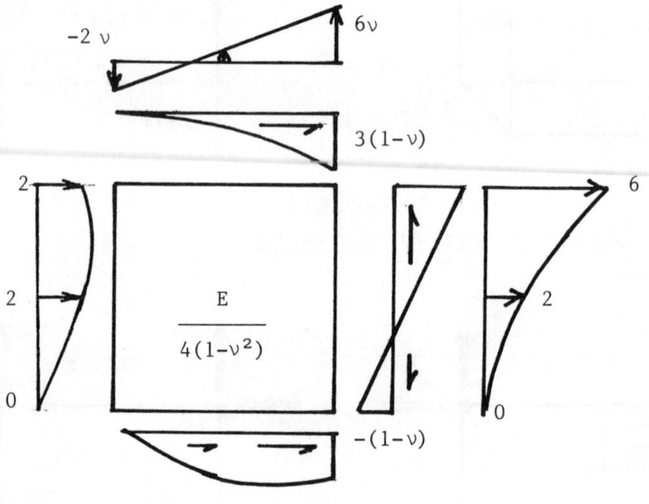

Boundary tractions

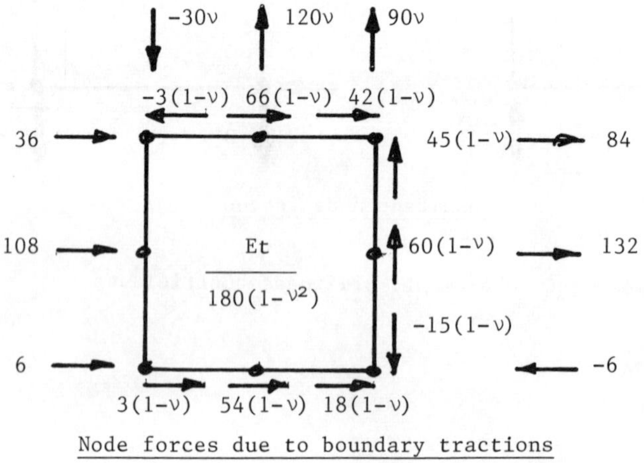

Node forces due to boundary tractions

Figure 7   8 noded square element: displacement of node 1

$$\{\bar{f}^e\}^T = \frac{-E}{4(1-\nu^2)} \; [(3+x + 2y-\nu(1+x)) \quad \frac{(1+\nu)}{2}(1+2x + 2y)]$$

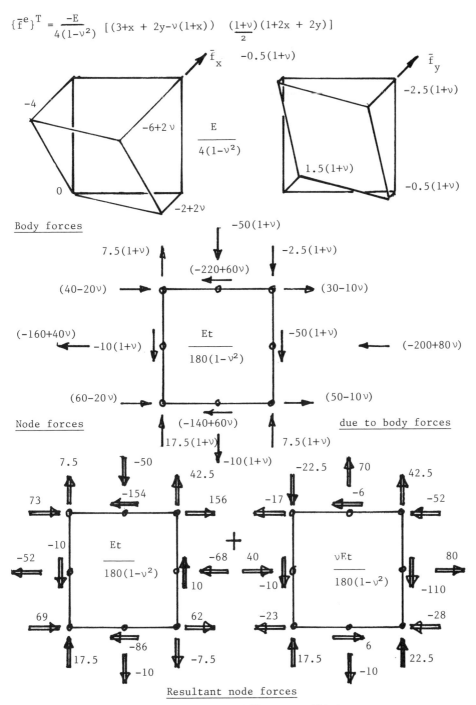

Figure 8   8 noded square element: stiffness coefficients

Of course this identity may be recognised as a virtual work equation, but appeal to the virtual work principle is unnecessary. Indeed formulations based instead on the 'identity', :

$$\{\Delta^e\}^T\{\bar{F}^e\} = \int [\bar{\epsilon}]^T \{\bar{\sigma}\}dv \qquad (21)$$

$$\text{external work} \qquad \text{internal work}$$

as a starting point may give a false sense of security to the student if he is unaware that he is imposing this identity in order to obtain an approximate solution. Since $\{\bar{F}^e\}$ is NOT in equilibrium with $\{\bar{\sigma}\}$, this 'identity' cannot follow from the virtual work principle.

ASSEMBLY OF A SYSTEM OF ELEMENTS

Interaction between elements is considered via nodal connections in a network of multi-terminal components. Generalised Kirchhoff's laws are applied to ensure compatibility of nodal displacements and equilibrium of equivalent nodal forces.

Considering nodes as free bodies, the nodal forces arise from :
a) interactions $-\bar{F}_i^e$ on node i from adjacent elements e. Denote the sum

$$\sum_e \{\bar{F}_i^e\} = \{\bar{F}_i\} \qquad (22)$$

The element forces $\bar{F}_i^e$ are dependent on element node displacements via the element stiffness matrix $[k^e]$. The sum $\{\bar{F}_i\}$ is the ith vector component of the product $[K]\{\Delta\}$ where $[K]$ is the stiffness matrix of the interconnected system of elements, and $\{\Delta\}$ is the nodal displacement vector of the system. Symbolically :

$$[K_{ij}] = \sum_e [k_{ij}^e] \quad \text{for elements connected in parallel} \qquad (23)$$

$$\text{(2x2)} \qquad \text{between nodes i and j.}$$

b) external loads or reactions. These are represented in a consistent way by dispersing load/reaction distributions on each element to act at the nodes. For a system of elements, dispersion to one node i can be expressed by a system shape function

$$N_i = \bigcup_e N_i^e \quad \text{for elements adjacent to node i} \qquad (24)$$

For compatible elements, the element shape functions must conform i.e. have uniquely defined values along element interfaces. In this case N_i has non-zero values only within elements and element edges adjacent to node i.

Furthermore the set of system shape functions must also have similar
properties as those for single elements i.e

$$\sum_i N_i = 1 \text{ and } \sum_i N_i \, r_i = r \tag{25}$$

and dispersed loads are statically equivalent to original loads.
Then the consistent load on node i :

$$\{F_i\} = \int [N_i]^T \{f\} dv + \sum \int_c [N_i]^T \{T_c\} da + \sum \int_b [N_i]^T \{T_b\} da \; ; \; [N_i] = \begin{bmatrix} N_i & 0 \\ 0 & N_i \end{bmatrix} \tag{26}$$

where the external forces take the form of body forces $\{f\}$, and line
tractions T_c and T_b which could be applied along element interfaces
and boundaries respectively. Due to the localised nature of N_i, for
compatible elements the only line integrals which are non-zero are those
for which the element edges are incident with node i.

Then in the stiffness method of analysis, the nodal equilibrium
equations :

$$-\{\bar{F}_i\} + \{F_i\} = 0 \tag{27}$$

are expressed as

$$[K]\{\Delta\} = \{\bar{F}\} = \{F\} \tag{28}$$

When some element edges have prescribed displacements, the
corresponding true reaction forces T_c or T_b are generally unknown.
However the number of equations can be reduced so that only those
corresponding to 'free' nodes remain.

WEIGHTED RESIDUAL EQUATIONS

Conforming elements imply compatibility of element deformations. Thus
of the three conditions for the true solution to an elastic problem –
namely compatibility, Hooke's law, and equilibrium – the first two are
satisfied but not generally the third in any strong sense. Lack of
equilibrium is defined in terms of residual body forces R_f, and residual
edge tractions R_c and R_b as follows :

$$R_f = f - \bar{f} \tag{29}$$

$$R_c = T_c - \bar{T}_c \; ; \; \bar{T}_c = \sum_e \bar{T}_c^e \text{ on element interfaces} \tag{30}$$

$$R_b = T_b - \bar{T}_b \text{ on boundaries} \tag{31}$$

where on the right hand side the unbarred quantities are specified loads or true reactions, and the barred quantities are derived as element responses to the nodal displacements calculated for the system.

The nodal equilibrium equations are now reconsidered in terms of these residuals. The derived force $\{\bar{F}_i\}$ at node i is the sum over adjacent elements :

$$\{\bar{F}_i\} = \sum_e \{\bar{F}_i^e\} = \sum_e \int [N_i^e]^T \{\bar{f}^e\} dv + \sum_e \int [N_i^e]^T \{\bar{T}^e\} da \qquad (32)$$

which for conforming elements reduces to :

$$\{\bar{F}_i\} = \int [N_i]^T \{\bar{f}\} dv + \sum_c \int [N_i]^T \{\bar{T}_c\} da + \sum_b \int [N_i]^T \{\bar{T}_b\} da \qquad (33)$$

using system shape functions. The equilibrium equation

$$\{F_i\} - \{\bar{F}_i\} = 0 \qquad (34)$$

can be expanded as :

$$\int [N_i]^T \{f - \bar{f}\} dv + \sum_c \int [N_i]^T \{T_c - \bar{T}_c\} da + \sum_b \int [N_i]^T \{T_b - \bar{T}_b\} da = 0$$

or as the weighted residual equation :

$$\int [N_i]^T \{R_f\} dv + \sum_c \int [N_i]^T \{R_c\} da + \sum_b \int [N_i]^T \{R_b\} da = 0 \qquad (35)$$

Thus a system equation is seen a posteriori to include all the different forms of residual forces. The normal Galerkin weighted residual formulation appears to consider a priori only the residual R_f.

For conforming elements, the system shape functions disperse the residuals to neighbouring nodes (i.e. nodes of one element or one edge) as consistent statically equivalent forces. Enforcing the nodal equilibrium equations puts the sum of the consistent residual forces to zero at a node. Hence the residual forces in their entirety are self balancing. However in the event of parts of the boundary having displacements specified, the true reactions and therefore the boundary residual forces are generally unknown. In this case the remaining known residual forces are not self balancing.

Residual forces themselves act as error indicators which can assist in judging the quality of a finite element solution. The solution obtained, if the model conforms to the displacement boundary conditions, is the the true solution for a modified loading comprising body forces $(f - R_f)$, and edge tractions $(T_c - R_c)$ and $(T_b - R_b)$. The significance of the residuals should then be judged against the inherent uncertainties involved in specifying distributions of loads.

Although convergence has not been proven, the importance of the shape function properties of linear completeness and conformity has been demonstrated. The engineering student should then appreciate that with a sequence of mesh refinements leading to closer proximity of the nodes, the residual forces will tend to fade in significance.

CONCLUSIONS

The aim of this paper had been to put forward a simpler, more direct formulation of the finite element method for structural models which is a valid alternative to the usual formulations. Hopefully it will be more readily understood by engineering students on first encountering the method - which may occur at various stages of an engineering course.

The use of the duality concept gives equal importance to the two roles of the shape functions in representing forces and displacements, and leads to the following advantages :
- the formulation has strong physical appeal without loss of rigour.
It generalises the simplest approach normally restricted to the constant strain triangle (9).
- single elements are considered first in a 'bottom up' formulation.
This is normally preceded by considering infinitesimal elements.
- the complete response of an element to displacements is required, and the integrals necessary to obtain stiffness coefficients are amenable to hand calculations.
- the stages at which approximations are introduced are clearly 'exposed'.
- the formulation leads naturally to accounting for all forms of 'load' residual, both within and between elements.

REFERENCES

1. Curtin, W.G., Qualitative analysis of structures. Report by a Sub-Committee of the Education and Examinations Committee, Institution of Structural Engineers, London, September 1989.

2. Brown, D.K., An Introduction to the Finite Element Method using BASIC programs. Surrey University Press, 2nd. edition, 1990.

3. Chandrupatla, T.R., Belegundu, A.D., Introduction to Finite Elements in Engineering, Prentice Hall Int. 1991

4. Robinson, J., Understanding Finite Element Stress Analysis, Robinson and Associates, Wimborne, Dorset, 1981.

5. Zhong, H.G., Beckers, P. Equilibrium default error estimators for
    the finite element solution. Report SA-139, L.T.A.S. Universite
    de Liege, January 1990.

6. Burnett, D.S., Finite Element Analysis from concepts to applications.
    Addison-Wesley, 1987.

7. Maunder, E.A.W., A direct formulation of finite element models in
    mechanics. Engineering Computations , 1989, 6, 248-258.

8. Frietas, J.A.T., Duality and symmetry in mixed integral methods
    of elastostatics. Int. J. Num. Meth. in Eng., 1989, 28, 1161-1179.

9. Livesley, R.K., Finite Elements: An Introduction for Engineers.
    Cambridge University Press, 1983.

AN OVERVIEW OF EDUCATION AND TRAINING IN FINITE ELEMENT ANALYSIS

RICHARD HENSHELL
Managing Director
PAFEC Limited
Strelley Hall, Strelley, Nottingham, NG8 6PE

ABSTRACT

The paper starts with a description of the differences between training requirements of different branches of Computer Aided Engineering and concludes that the requirements for FE rather more stringent than for other branches. Following a description of the more detailed requirements for finite element analysts and developers there is a list of the engineering science training required. It is generally concluded that the engineering science requirements for FE are quite demanding. Equally demanding is the need for engineering intuition.

INTRODUCTION

I was very pleased to be asked to make this presentation and I chose the title for it. I found out later that Professor Owen will be speaking on the first day of the meeting and so I have no wish to duplicate the important messages which someone so eminent has had for you.

My own claim to be here is, I suppose, that I have been involved with finite elements for 27 years; I did research in the subject for many years followed by a period of teaching and during this time I ran many courses on the theoretical and practical aspects of the finite elements. Following that I started PAFEC Limited which was initially only interested in finite elements but latterly in many other branches of computer aided engineering. So I can now claim to represent both the academic and industrial aspects of finite element use as part of the wider subject of Computer Aided Engineering (CAE).

EDUCATION AND TRAINING IN OTHER BRANCHES OF CAE

PAFEC has been involved with many parts of Computer Aided Engineering. The education and training needs of the branches are naturally different. With computer aided draughting, the main teaching emphasis is on the interactive use of a computer, the mastery of a subset of the system's capabilities and a familiarity with the documentation to find out how other facilities are used. Computer aided manufacturing is a little different: the subject involves many minutiae and these have to be related to the functions of a program in some detail.

Training in the use of computer aided process planning or manufacturing decision systems requires knowledge about how to express, in a computer interpretable form, the decisions which a person takes for granted.

None of these seems to be as difficult as the problem of training the finite element practitioner.

Finite element analysis has passed through the early phases when pioneers were relatively few, very dedicated and not easy to understand by the general engineering community. Self training was adequate in those days. Professor Zienkiewicz had a great impact in those early days in popularising the method and in causing many of us to start working in the field. When a method is in its infancy it needs an enthusiastic attitude and often a rigorous attitude is inappropriate.

THE FUNDAMENTAL BASIS FOR FE

At that time and probably ever since then there have been at least two views of the theoretical basis for the FEM.

1. The first and simplest to grasp for the engineering student is to say that a structure is divided into elements and that for each element we derive a stiffness matrix which relates the forces at the node of an element to the generalised displacements of the element. Equilibrium balances between elements give a system stiffness matrix. It is natural to consider the elements as rather complex springs.

2. The second and more rigorous view is to present the FEM as being the result of expressing the displaced shape of a structure in terms of some assumed displacements at points called nodes. The energy of the structure is written down for the whole structure and we use some minimum energy principle to equate the differentials of the energy to zero. This gives us, in general, exactly the same sort of equations as we get from the method 1. above.

We always used method 1. to explain the method to newcomers and gradually changed to method 2. We felt that a better physical understanding was achieved by doing this at the early stages.

But naturally, there are limitations to the method 1. and if we proceed with it too long, confusion can occur.

The concept of applying forces at nodes does not seem strange to the novice. But when the student thinks carefully he can see that the force transfer between two dimensional elements in a plane occurs right along the adjacent sides of elements and not just at the nodes.

At some point we shall have to consider various energy formulations. The first one introduced will normally be the principle of minimum potential energy calculated on the basis of displacement assumptions. Some of the other principles are rather difficult mathematically and would probably never be considered unless the student is likely to develop FE programs. The figure 1. shows the relationship between the different variational principles based upon energy.

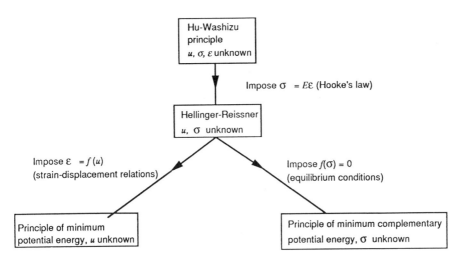

Figure 1 Connection between variational principles

I remember a distinguished US professor asked me about 20 years ago to state the variational principle which applied to a particular situation. I had never thought about it and never remembered seeing one in the literature. He said you should never do any calculation without knowing the variational principle. In practice everyone who had done that type of calculation based them upon the principle 1. above, suitably restated to relate to the case in point. Thinking about the case later I found a variational principle which fitted the bill. The case concerns transient temperature (T) calculation in a solid solved by time marching. For the sake of simplicity I will consider uniform conductivity (k), specific heat (C_p) and density (ρ). The partial differential equations relating to this situation is well known

$$\nabla^2 T = \frac{C_p \rho}{k} \dot{T}$$

But appropriate variational forms are not widely known. Some possibilities are

$$\int_{volume} \left(\nabla^2 T - \frac{C_p \rho}{k} \dot{T} \right) T \, d(volume) = minimum$$

where $\nabla^2 T$ and $\dot{T}$ are taken as constant

$$\int_{volume} \left(\nabla T . \nabla T - \frac{C_p \rho}{k} \dot{T}.T \right) d(volume) = minimum$$

where $\dot{T}$ is taken as a fixed but unknown vector for a particular time step

The contorted nature of the constraints shows how unnatural this process is and obviously the cruder method based upon rather Victorian principles may be more appropriate.

This situation is unsatisfactory and naturally demands a more rigorous treatment so we can now offer two more principles for the basis of the FEM.

3. The principal of virtual work - this tends to have been favoured by civil engineers and doesn't lead to a very natural approach outside solid mechanics.

4. Galerkin's method or the weighted residual method in which the partial differential equation is multiplied by shape functions. Again this approach is not very natural to the student but has the advantage of easy application in almost all situations. It seems to be the best basis if method 2. cannot be applied.

In summary then, methods 1,2 and 4 allow a natural progression of complexity and rigour.

Naturally the finite element program developer needs to consider this subject of the theoretical basis quite carefully. But he will generally find that no rigorous theory leading to such things as convergence proofs can be constructed for the new and difficult situations (for example convected non-linear heat transfer). In such cases, the developer gives up on theory and trusts to his judgement and heuristics.

From the viewpoint of education and training we therefore have few rules that we can give to the student. We have to do at least four things:-

1. Expose the student to the theoretical background at a level appropriate to his abilities.

2. Give the student experience so that he knows what strategies are viable, and what accuracies can be obtained, with a given strategy, what he should question and how he should make checks.

3. Provide checks in FE programs so that poor data cannot go past the input phases. There will be lots of areas where a checking system cannot know definitely whether a particular data entry is wrong. But at least if it is questionable a warning can be given.

4. Checks on output such as force balances, stress discontinuities are very useful both for providing limits on the accuracy of an analysis and also for educating the user for future analyses.

Perhaps the most important point to be made to the student on this subject is that the finite element method is an approximate method which tends to give more accurate results as the mesh sizes, load steps, time steps etc., decrease.

TRAINING IN IDEALISATION

Given a particular problem to analyse, the FE user has to replace this by an FE model. The density of mesh is an important issue which I shall consider later under the heading of convergence. For the present, I shall just consider issues like:-

1. When does a plate or shell problem change from being thin to moderately thick to being a full three dimensional problem?

2. When does an open section beam change from being a beam finite element to a folded plate structure?

3. When can a hole, boss, cut-out etc., be ignored?

4. When can an almost axisymmetric structure be considered as truly axisymmetric?

These are all questions of judgement and to exercise that, the trainee or student needs both an understanding of the way in which various types of finite elements work and just as important, a flair for the behaviour of engineering structures.

In an environment with some experienced staff close by, the trainee has few problems. But if there is only one raw analyst in a company it is difficult. At one time PAFEC Limited produced a short introductory manual 'PAFEC GET STARTED' to help solve this problem. One double page spread in that document gave the user some hints about how to do an analysis. It still has value to the novice in the early stages, but a complete map of the capabilities of the software today would need a tennis court. The form used in the late 1970s and taken from PAFEC 75 Theory Results is shown in figure 2.

CONVERGENCE OF THE FEM

The engineering student will usually accept that the FEM will give exact answers as the finite elements become vanishingly small. Proving that to be so is another matter. Proofs of convergence now exist for some cases; those proofs are generally capable of identifying, by default, cases in which convergence is known not to occur. But there is a simple patch test which renders the convergence proofs rather obsolete. In practice there doesn't seem much point in exposing the finite element practitioner to these methods although they are very important indeed to the developer.

BOUNDARY CONDITIONS AND LOADS

This is an area where new users of the FEM often go wrong. As in the case of the idealisation issue, the student needs an engineering understanding of how his structure will behave and a knowledge of what the various clamps, slides, pin joints etc., which he can describe in his data actually do. There really is no alternative to this combination of engineering feel and education in the mathematics of the process.

As a general observation though, new FE users seem to me to expect the boundary fixities to be stronger than is actually the case in a real engineering structure.

Likewise, the application of loads has to be done with care. Very often there will have to be guessing to arrive at loads. Engineering judgement is needed for that. The actual application of the loads is not usually difficult.

DE-IDEALISATION

Having done an FE analysis, the results flash on a screen or pour out onto printer paper. If approximations had to be made at the stage of idealisation then the results have to be viewed with those approximations being borne in mind.

For example, if a particular geometric feature has been omitted from the model, it would be necessary at the de-idealisation stage to realise that effects of this geometric approximation will be in the results.

Decision chart for finite element methods

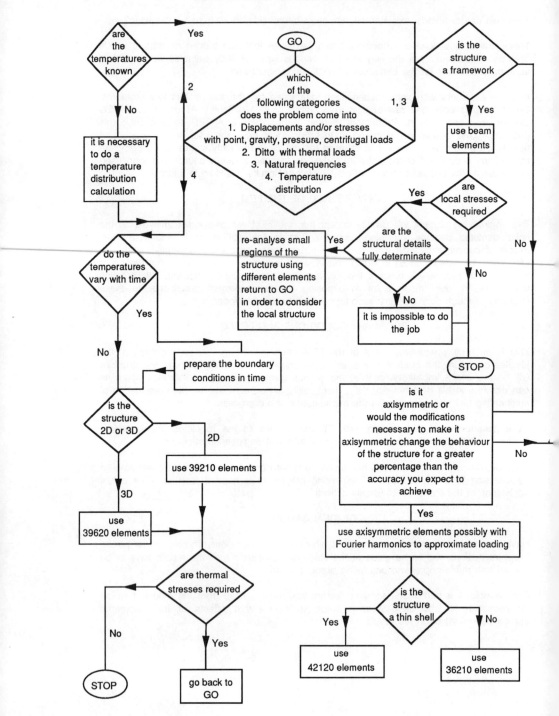

125

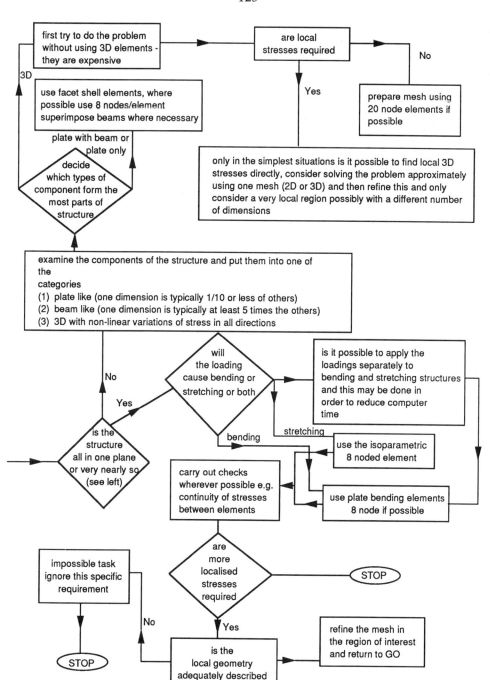

first try to do the problem without using 3D elements - they are expensive

3D

use facet shell elements, where possible use 8 nodes/element superimpose beams where necessary

plate with beam or plate only

decide which types of component form the most parts of structure

are local stresses required

No

Yes

prepare mesh using 20 node elements if possible

only in the simplest situations is it possible to find local 3D stresses directly, consider solving the problem approximately using one mesh (2D or 3D) and then refine this and only consider a very local region possibly with a different number of dimensions

examine the components of the structure and put them into one of the categories
(1) plate like (one dimension is typically 1/10 or less of others)
(2) beam like (one dimension is typically at least 5 times the others)
(3) 3D with non-linear variations of stress in all directions

will the loading cause bending or stretching or both

No

Yes

is the structure all in one plane or very nearly so (see left)

is it possible to apply the loadings separately to bending and stretching structures and this may be done in order to reduce computer time

bending

stretching

use the isoparametric 8 noded element

carry out checks wherever possible e.g. continuity of stresses between elements

use plate bending elements 8 node if possible

impossible task ignore this specific requirement

are more localised stresses required

STOP

STOP

No

Yes

is the local geometry adequately described

refine the mesh in the region of interest and return to GO

If the type of finite element used or FE model is cruder than some geometric feature, then the analyst needs to assess the effects of using a better class of element or refined mesh.

De-idealisation requires again much engineering judgement and also a knowledge of how the different finite elements operate. There really is no alternative to these types of knowledge - no computer program could adequately cater for them.

DYNAMICS

In the particular case of dynamic problems, in which the most common analysis involves the calculation of natural frequencies of the structures, this is usually no more difficult than the problem of general stress analysis. In fact it can be easier because the mesh densities required to give a particular accuracy maybe less.

The remarks made about fixity apply even more importantly in the case of dynamics. Nothing in the world has a clamp as strong as the mathematical condition implies. The new user almost always truncates his problem too early and assumes too great a rigidity at the boundaries. Very often this can be overcome with a very crude model of the next level of structure.

The dynamic problem can be extended into transient response, harmonic or sinusoidal response, acoustic analysis, random vibrations. None of these are likely to present very difficult problems but they should probably not be tackled by the new finite element user until experience on other problems has been obtained.

THERMAL PROBLEMS

It is rather common to carry out a transient temperature calculation and then to follow this with a calculation of the stresses which are caused at different times during the thermal transient. The ideal is to do the two analyses with one mesh. This requires rather careful choice by the analyst who needs to choose small elements in the direction of a thermal shock and small elements in the area of stress concentrations. A time step has to be chosen for the thermal transient part of the analysis and this must be carefully matched to the size of the elements in the direction of any thermal shock.

NON-LINEAR STATIC ANALYSIS

Not surprisingly, non-linear static stress analysis requires a greater degree of education and training than any linear calculation.

One of the first things which needs to be taught is the question of whether and when a non-linear analysis is appropriate. Generally, a non-linear analysis is much more expensive than a linear one and so it should not be undertaken unless a linear one is demonstrably inappropriate.

The most common limitation in accuracy with non-linear analyses is the accuracy with which material properties are known. There is no alternative for engineering experience to assist the analyst in judging whether the numbers to describe materials are accurate or not.

Apart from the material parameter difficulty, there is the question of the method of using a non-linear code. This is always much more difficult than a linear one and the more non-linear your problem, the more difficult is the use of a program.

The analyses of such difficult problems as the gross distortions caused by crashes and fragmentation as a result of impact represent the most difficult class of problems and are definitely not to be under-taken by the under-trained or faint-hearted.

TRAINING IN QUALITY

When a new science is in its infancy and few people are involved, quality is seen as an irrelevance and may even be deleterious to progress.

But as a science matures, more people become involved and actions taken become more safety critical. Therefore an increasing interest in quality is required.

Today, the finite element method is mature and the analyst is expected to keep proper records of all he does, how he has arrived at decisions, the sources of his information and to carry out proper and thorough checks on what he has done. He needs to be trained in these processes. During third level education, the student probably finds subjects such as these rather dull and therefore is not likely to be very attentive. My own view is that quality in FE is best taught after the student has obtained his qualification and has started his first job.

TRAINING IN THE USE OF GRAPHICS

Graphics pre- and post- processors are now common place and naturally one must question the level of education and training needed to use these appropriately. In general, not much detailed training is needed: the systems should always have the capability of being almost self-taught and self-prompting. If this is not the case, then there is something wrong with the program.

128

ESSENTIAL ENGINEERING SCIENCE FOR THE FINITE ELEMENT ANALYSIS

One of the continuing themes through this paper has been the need for a proper grounding in engineering science.

I list below some of the subjects which I see as essential for almost all analysts and later the subjects needed for some specialisms.

Essential for all analysts:

Concept of displacements, stress, strain, temperature etc., as field variables

Relationships between the above qualities

Some energy principles

Poisson's ratio effects

Saint Venant's principle

Concept of field variables satisfying partial differential equations

Understanding of boundary conditions

Foundations of thin plate and moderately thick plate theories

Principal directions and moments of areas of cross-sections of beams

Equilibrium equations

Principle of superposition in linear analysis

Concepts of continuity of functions in many dimensions

Intuitive feel for structural behaviour

Limits of linearity of materials

Difference between small and large displacement theories

Dynamics of elastic structures, natural frequencies, mode shapes, resonance, response

Some needs for more specialised areas:

Shell theories

Large displacement theory

Large strain theories

Plastic flow theories

Creep, viscoelastic and viscoplastic analysis

Analysis of infinite and infinitesimal regions - crack analysis

Gyroscopic forces

Fatigue analysis

Fourier analysis for axisymmetric structures

Transient, sinusoidal and random vibration

Various damping types and theoretical treatment

Acoustic loading on structures

Wave equation for acoustics

Fluid loading on structures

Different types of composite material e.g. uni-directional, weave etc., and their characteristic material properties

Failure theories for composite materials

Coupling of effects - e.g. piezoelectricity

Convection - natural, forced and the corresponding heat transfer coefficients

Analysis of constrained layer materials, sandwiches

Numerical analysis applied to iterations, marching, Newton Raphson

CONCLUSION

The general theme which has been running through almost all the sections so far is that the finite element method requires a good deal of engineering understanding to get anything worthwhile out of it. One of the implications of this is that no matter how easy finite element programs become to use we shall still need trained engineers. In fact, as programs become more complex and can solve more difficult problems the level of general education required for engineers will increase.

There is separately a need for understanding how the finite element program actually works. This gives insight into the limitations of particular classes of finite elements or types of calculations in order that the engineering judgement can be used appropriately.

There is constant pressure on finite element developers to make their programs easier to use so that less detailed knowledge is necessary. That market pressure will always exist and therefore developers will certainly pay heed to it. But, the dream of being able to analyse something of great complexity by pressing a button from a CAD model is not likely to be realisable.

Thus the drive for user friendly software is somewhat at odds to the needs of education and training because important engineering knowledge is being locked up in programs and hidden from the user in detail and appearing in the form of increasingly robust FE systems. But we are a long way from being able to use these powerful programs without proper understanding.

AN INDUSTRIALLY RUN FINITE ELEMENT TRAINING COURSE

A.N.Pickering and D.C.Ricketts
GEC–ALSTHOM Engineering Research Centre,
Cambridge Road, Whetstone, Leicester LE8 3LH

ABSTRACT

A finite element course "The Practical Use of Finite Elements" has been run by the Engineering Research Centre for some time. The course, of three days duration, is aimed at new practitioners in the finite element field. The course participants are drawn from industry, in the main from GEC companies, and have a widely varying background. This paper describes the course lecture content and structure. Since the aim of the course is to give practical guidance in the use of the Finite Element Method with a minimum of mathematical theory, the hands–on sessions are considered a vital ingredient. Their value as a teaching mechanism is discussed here.

INTRODUCTION

The use of Finite Element Techniques in structural and other types of analysis has increased dramatically over the past few years, as the availability of computer power has improved and become less expensive. This has broadened the range of personnel who wish to use the technique, and this more widespread use requires careful handling if costly errors or misjudgements are to be avoided.

To help the process of making the Finite Element technique more widely available, and its use more reliable, the Engineering Research Centre (ERC), part of GEC–ALSTHOM, has been running training courses for engineers who are shortly to be involved with the use of Finite Elements in the execution of their work.

The course attendees have had a varied background, ranging from many years experience in traditional engineering analysis, to graduates fresh from university with little practical experience. In addition, the computer literacy of the students has varied at least as much as their engineering experience, although in this case generally in favour of the younger graduate.

ERC's Finite Element activities, which have been carried out for over twenty years, can be considered to take two forms:

- Use of finite element programs in engineering analysis.
- Development and support of Finite Element systems.

The first of these activities is carried out on a contract basis; the code that is used varies depending on the application. It has been noted that as time goes by the type of work which has been subcontracted by clients has changed from two dimensional linear elastic to either nonlinear analysis or geometrically complex linear three dimensional problems. This is due to the increasing development of in-house capability by our clients.

As part of ERC's development activity, two structural analysis codes have been produced; MELISSA and MELINA. These continue to be enhanced at ERC, and their use is supported within a number of GEC companies. The development of good analysis practices and their adoption by GEC product units is a further continuing area of interest.

The company has also been involved in the NAFEMS initiative, and several NAFEMS contracts have been awarded to ERC. Ongoing studies have contributed to the benchmark exercises for linear elastic problems, while currently the NAFEMS Workbook of Examples is being compiled by ERC.

COURSE REQUIREMENT

Shortfalls in knowledge can be difficult to recognise, and this is particularly so with computer aided techniques which produce answers of apparent high accuracy. Through the widespread use of Finite Element techniques over many years ERC engineers have developed a high degree of skill in their application. ERC's involvement in software development and supply brought the key finite element engineers of ERC into contact with new finite element users in the product companies, and it became clear that although in many cases, the computer facilities afforded by these companies were better than our own, the engineers were not always sufficiently skilled to make the best use of them. In fact, in many ways the high performance of the software and hardware available lured the novice user into the utopian "black-box" analysis arena. The ease with which a large model "appearing" to represent the component can be created can only be contrasted with the subsequent difficulties encountered trying to debug the model, perform the analysis and make sense of the output.

It was clear that ERC had skills and experience which could, through training, be transferred to the new users of the Finite Element technique. The course "The Practical Use of Finite Elements" was therefore produced, and conveniently the timing enabled the new NAFEMS text books ("A Finite Element Primer" and "Guidelines to Finite Element Practice") to be used as the core material. The emphasis of the course was very definitely on practical use, with only a minimal theory content to provide the background required. The scope was firmly restricted to the finite element user, rather than the finite element methods developer, and much of the content was based on experience, good practice and the promotion of thought and organisation into a discipline with the capability for being swamped by over-automated software tools. The fundamental message we were attempting to promote was that a bad analysis is often worse than no analysis at all.

COURSE OVERVIEW

The course is run over three days and is split approximately 50:50 between lectured material, and practical "hands-on" sessions designed to illustrate the principles of the technique and the fundamental nature of Finite Element software.

The format is such that subjects dealt with in the lectures are illustrated first by discussion of example analyses, and subsequently through simple benchmark problems posed in the hands-on sessions.

The overall program is as follows:

- Day I

    Basis of Finite Elements

    Developing a Finite Element Model

- Day II

    Results Processing

    Hands-On Exercises

    Dynamic Analysis

- Day III

    Modelling of Structural Details

    Hands-On Exercises

    Other Types of Analysis

The emphasis throughout is on linear structural analysis since this is where the technique is most widely applied. Nonlinear analysis and the solution of other field problems (e.g. thermal analysis) are only briefly discussed at the end of the course, but this is not seen as a deficiency since the underlying principles presented and the analysis approach promoted are applicable to all forms of Finite Element analysis.

LECTURE   CONTENT

Basis Of Finite Elements

This lecture provides a broad insight into the underlying theory of the Finite Element method, and introduces those concepts necessary for the subsequent material to be understood. The fundamental relationships of equilibrium and compatibility and their connection through the material stress–strain law are described. The three relationships are developed for a simple bar structure, and are combined to produce the displacement equation. The components of this equation are identified as the unknown displacement (r), the known applied force (R) and the structural stiffness matrix (k = AE/l). The application of the Principal of Virtual Displacements to the same problem, and the delivery of the equations of equilibrium from simply equating internal and external virtual work give the basis for the development of the Finite Element equations from an assembly of individual element stiffness matrices. The matrices formed at each stage of the element stiffness calculation and assembly are described, with particular reference to the Jacobian matrix.

The basic types of elements and their formulation are described briefly with examples of where their use is most appropriate. Generic element types discussed are :

- Bars (or trusses).
- Beams.
- Two Dimensional solids for plane stress, plane strain and axisymmetric problems.
- Three Dimensional solids.
- Plates
- Shells

The approximations resulting from the fundamental assumptions are summarised; in particular, the discontinuous nature of stresses across element boundaries. Since the element stresses are calculated from the displacements within the element, the interpolation of these from the nodal displacements using shape functions is dealt with in some detail.

Developing a Finite Element Model

The development of the Finite Element model is presented as a complex task involving several stages. It is emphasised that these stages must be present, in the order stated, to ensure that the analysis is performed in an organised, traceable manner, which should eliminate errors and ultimately increase the analyst's confidence. By performing analyses in the prescribed manner, and documenting the process, greater reliance on Finite Element results can be achieved, and the full cost benefits of employing the technique can be realised. The analysis process stages are described in the following sections.

Pre–Analysis : The careful consideration of all aspects of the analysis to be performed is encouraged at this early stage. As a minimum, the following questions should be addressed:

· What is required of the analysis ?
· How much of the structure is to be modelled ?
· How much detail is required ?
· What analysis system should be used ?
· Is a benchmark required ?

In addition, it is stressed that the pre–analysis stage is when all hand calculations should be performed, before the Finite Element results are available.

Boundary Conditions : The different types of boundary condition available to the Finite Element analyst are described with examples, where appropriate, of the physical situations they may represent. It is stressed that in order to achieve the correct behaviour, the boundary conditions applied must correspond with the element types used. The correct suppression of rigid body displacements, and the effect of failing to do so is also presented. The advantages of recognising and exploiting simple symmetries are shown.

Loading : The various types of mechanical loading available within most current programs are described, but it is pointed out that, since the load vector contains only a single value for each degree of freedom, all forms of mechanical load (however complex they may appear) are converted to point loads at node points. The user is warned against applying localised loads as point loads, since these would theoretically produce infinite stresses. Preferred methods for applying such concentrated loads are discussed, as are the assumptions of die–away length made when deciding on the method of load application and the mesh requirements adjacent to the point of load application. As part of this lecture the way in which the Finite Element program converts distributed loads to point values is described, and the importance of kinematically applied loads is underlined. The conversion of node point temperatures to thermal strains, and then to nodal point loads for use in a thermal stress analysis is also described.

Material Properties : The material properties required for different types of analysis are identified, and the reason for their inclusion described. Most of the lecture on material description is devoted to the simplifying assumptions made regarding material nonlinearity. It is emphasised that many material properties vary with temperature, strain, time or all three. The importance of making the correct assumptions for the material behaviour is stressed, as is the extra complexity of the nonlinear analysis required for complex material models.

Model Development : The model development phase of the analysis process incorporates all of the decisions made in the previous stages. The students are encouraged to address the following in the mesh definition activity :

· Take account of all boundary conditions, loads and material boundaries.

· Anticipate load paths and high stress gradient areas.

· Coarse preliminary models may be run if the problem is not well understood.

· Use sensible mesh refinement techniques to optimise the overall model size without introducing errors due to inadequate mesh detail or badly distorted elements in critical regions.

· Select the most appropriate element types, taking care when combining element types.

· Use all means possible to optimise the solution procedure, through equation re–ordering or substructuring techniques.

The use of these guidelines is then illustrated through examples, during which several other effects and techniques are introduced. Among these are the importance of geometrical accuracy when modelling curved surfaces, comparisons of the performance of low and high order elements, the use of multi–point constraints and the effects of reduced integration. The conclusion to the model development phase is a description of the ways in which the model integrity can be verified using modern graphics tools to highlight vacancies, duplicate elements, distortion and lack of connection between submodels.

Results Processing

The types of results processing promoted by the course fall into three categories: Model integrity checks; Solution diagnostic checks; and result appraisal.

Model Integrity Checks : The student is made aware of the types of checks that can be made, which can give confidence in the results. The balancing of loads and reactions is one indication of successful program execution, and cursory checks on surface normal stresses should confirm correct application of loads. More detailed checks on localised unaveraged stress discontinuities and comparisons of the overall behaviour with that predicted by hand calculations or test data are also discussed.

Solution Diagnostics : The types of diagnostics which should be available following the solution are described. These include diagonal decay, ill-conditioning and element shape diagnostics. The practice of assessing these parameters prior to detailed assessment of the results is encouraged, and the importance of recording warnings, errors and the justification for their treatment is stressed.

Result Appraisal : The teaching of detailed results appraisal includes some of the earlier global checks, but in finer detail. The difficulty of predicting result quality is illustrated with a distorted mesh example, where the maximum errors (defined using stress discontinuity) often occur away from the apparently worst mesh regions. This emphasises the need to investigate the quality of the results without any preconceptions. As part of the detailed appraisal, the re-assessment of the modelling assumptions must be made, in particular with reference to linearity. Linear analyses producing large deflections or stresses exceeding the material yield stress are indicators that the original assumptions may be invalid.

Dynamic Analysis

The use of Finite Element methods in the analysis of dynamic problems is dealt with briefly, and although the full dynamic Finite Element equations are developed, these are reduced to the undamped free vibration case for the detailed discussion. Since the solution of this Eigenvalue problem involves iteration, the importance of choosing appropriate solution algorithms and mesh densities is stressed. In addition, the use of the Guyan Reduction technique is described, and a large part of the lecture is concerned with the correct choice of master degrees of freedom for the reduced problem. The effects of poor choice are illustrated using a simple example with many different sets of master freedoms, each producing different modes.

Modelling Structural Details

The final full lecture of the course addresses some of the common problems facing the Finite Element engineer; the inclusion or exclusion of a structural detail (e.g. a hole), the modelling of a complex interaction between components, the use of symmetry to reduce analysis size and cost and the smearing of properties to reduce modelling effort. These subjects are discussed, and proven solutions are presented. By this stage in the course, the delegates usually have sufficient background to be able to discuss their own products in Finite Element terms and the session usually closes with the students posing problems sufficiently diverse to keep the lecturer on his toes.

Other Types of Analysis

The course predominantly deals with linear stress and dynamic analysis, but the principles are applicable to many similar field problems, and of course to nonlinear problems. The course is closed with a brief summary of the types of non–structural problems and nonlinear structural problems which can be addressed, and in the latter case, the extra complexity which is introduced both in the solution cost and in the higher level of understanding required to model nonlinearities effectively.

HANDS ON SESSIONS

The hands–on sessions are interspersed with the lectures and are designed to enable course attendees to learn the practicalities of running a finite element code. The course participants are split into groups of two or three, and each is allocated a terminal and computer user identification. They are given a number of finite element problems to solve: the amount of data input necessary varies from none in the case of the first example which has both geometry and loading fully described and in a form which enables the problem to be run immediately to later examples where the loading has to be entered or the mesh altered in some manner. It is at this point that the mix of computer literacy becomes a problem as files needed to be manipulated. Thus, wherever possible the groups are chosen to provide each group with someone who is familiar with the operating system or something similar. (A recent innovation on the course is an optional computer familiarisation session prior to the commencement of the main body of the course at which file handling and manipulation are taught.) The examples enable such things as rigid body motions (lack of fixity), the effect of kinematically equivalent and statically equivalent loadings, element distortion and linearity to be demonstrated. The greater part of these examples have been obtained from the

NAFEMS linear elastic benchmark activities, as these are usually relatively simple in geometry and loading and have a well documented solution. Supervision of the hands-on sessions is carried out by ERC staff other than the course lecturers in order to provide everyone involved with a break. The experience gathered from the hands-on sessions has been most useful in carrying out a contract for NAFEMS for the writing of a Workbook of Examples to accompany the NAFEMS proposed thirty hour finite element syllabus. This document, when completed is designed to be used as an exercise book for the necessary hands-on work for the NAFEMS course. This work is envisaged to be carried out, under the guidance of a mentor, after the conclusion of the course. The workbook, it is envisaged, will have a number of examples, each laid out in the form of a problem specification which describes the geometry and loading, and an analysis specification that describes alterations to the model and/or loadings. Each example will have a commentary describing the salient features of the analyses and the target solution which ought to be obtained.

SAMPLE COURSE EXAMPLE

Throughout the course examples are used to illustrate the points raised in the lectures. One such example is the "Plate With a Circular Hole" which is the classical stress concentration factor problem. This example is referred to at several points within the course, since it demonstrates many of the key issues, in particular :

- Constraints required to take advantage of symmetry.
- Importance of geometrical accuracy close to regions of high stress gradient.
- The effect of misplaced midside nodes, particularly on curved surfaces.
- The care required when using preprocessing tools to develop curved surface models.
- The effect of different element order and mesh density on the solution.
- The use of multi-point constraint equations in mesh refinement.
- The use of distorted elements between coarse and fine mesh regions.
- The concept of die-away length and its importance.

Figure 1 shows one sample mesh used in the discussion of this problem.

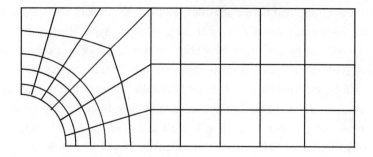

Figure 1 . Example Problem – Plate With a Circular Hole.

COURSE HISTORY

The course was developed in 1987, and has been run three times a year at Whetstone. Initially, the course was publicised within GEC by use of internal newsletters which had a mechanical engineering readership. Subsequently, the course has also been marketed by use of an insert within NAFEMS publications. Thus approximately 80 students have been on the course. The course has also been run at client sites, both in the United Kingdom and Europe. This has the benefit of being a more cost effective method of training a number of students, with the further advantage that the computer system and finite element code are the customer's own. This helps the student gain maximum benefit from the hands–on sessions. The site course has been run a total of 7 times for 4 client companies. By this means a further 70 students have been trained.

At the end of the course the participants are asked to fill in a questionnaire to assess the course content, style and presentation. These are then reviewed to establish whether the course needs alteration. The response has been extremely positive in nearly all cases the only exceptions to this having been students with specialist interests in forms of analysis which are non–structural. They have felt that the bias towards stress analysis has been too great. The amount of repeat business both for the Whetstone course and customer site courses has been very encouraging, and this is a very good test of customer satisfaction.

CONCLUSIONS & FUTURE

It is clear that there is a requirement for training in the use of the finite element method, particularly with the continuing real reduction in the cost of computing power. The course, as presently designed, seems to have been greeted favourably, particularly if repeat business is considered. However, in the not–too–distant future a number of decisions will need to be made.

The major one that is faced is whether the course should be upgraded to include the necessary material to meet the NAFEMS thirty hour course syllabus. If this is in fact done this will have a number of implications; most importantly, the course will have to be extended to one week duration. This might well prove to be a disincentive to potential clients as their personnel will need to be on the course longer and hence 'non–productive' for longer. Additionally the cost of the course would need to be increased pro–rata. If the decision is made to enter the course for approval by NAFEMS it is not clear how the post training course mentor system could be implemented in those companies which are just introducing the finite element method for the first time. ERC could provide this extra assistance, but again the cost implications to the client may prove prohibitive.

FINITE ELEMENT TRAINING AND EDUCATION - A SOFTWARE VENDOR'S PERSPECTIVE

DR. GARY C. CARTER
Co-ordinator, PDA Institute of Technology
PDA Engineering International Ltd.
Rowan House, Woodlands Business Village, Coronation Road,
Basingstoke, Hampshire RG21 2JX, UK

ABSTRACT

PDA Engineering is an international leader in the development and application of Mechanical Computer-Aided Engineering and Materials Technology. The company offers not only engineering software but also engineering consultancy. This, together with many years of working with and training of engineers in a wide variety of industries, gives PDA a unique understanding and direct experience of the training and educational needs of the engineer working in finite element analysis.

PDA have for many years given courses in the use of the PATRAN engineering software system. Topics covered include element verification, mesh optimisation and selection of appropriate element types. Our training methods are biased towards the practical use of finite element analysis, together with an explanation of the underlying method rather than the fundamental mathematics and theory. There is no shortage of good courses in the general theory of the method but the 'art' of applying this knowledge to the task of creating a quality assured finite element model is not something which is traditionally taught. This paper will discuss some of the issues involved in teaching the 'art' of finite element modelling.

INTRODUCTION AND MOTIVATION

As a company, PDA Engineering work with engineers who perform a wide variety of analyses; structural analysis, fluid dynamics, thermal analysis and durability assessment to cite just a few. Given this broad range of interests, the traditional system of teaching the theory of the finite element method in terms of structural mechanics applications is not always appropriate. Indeed coming from a background of computational fluid dynamics I have first hand experience of the problems this can cause. Drawing on this experience, I was involved in the development of an introductory course in finite elements [1] aimed at engineers, mathematicians and physicists who did not necessarily have any structural analysis experience.

PDA Engineering's close working relationship with our customers has lead us to recognise a gap in the knowledge required by an engineer in order to produce a good finite element analyst. This is specifically in the area of modelling.

It sometimes comes as a shock to an engineering manager that on return from a finite element training course, his engineers are still not sure how to approach the finite element discretisation and loading of a model. This is certainly our experience, not only as practicing engineers but also through our contacts with engineers attending PDA for software specific training. In response to the increasing demand for training in finite element modelling we are now providing a course in 'The Art of Finite Element Modelling' [2].

This demand is coming from various sources; the finite element analyst, the engineering manager and more recently from the designer. With many Computer Aided Design systems now offering a limited form of finite element analysis, there is a new demand for training concentrating on achieving the best results from these systems. This paper will look at some of the areas which traditional training methods do not cover and discuss ways in which these techniques can be successfully taught.

There are benefits in providing training of this nature both for the software vendor and the user. The user becomes more efficient and effective in his use of the software. This not only enhances the image of the software to the credit of the vendor but also reduces the level of on-going support that the vendor needs to supply. The user will also see improved productivity, reduced costs and a more accountable quality assurance process. There are also significant benefits in engineers using external sources to obtain this training. It would be a reasonable assumption that the best finite element engineers would be engaged in providing the in-house training. To do this however means removing these people from production work. Furthermore, the use of in-house trainers does not provide as many opportunities for introducing new ideas and methodologies as does the use of an independent training organisation.

Up until very recently, the two main sources of information available on the finite element method were the many excellent textbooks on the subject, [3] - [5] (to cite but a few), and the documentation for the commercial programs used for the majority of finite element analyses performed today. However, the former have tended to concentrate on the theory and mathematics of the method whilst the latter have inevitably been biased towards documenting the format of the input data for a particular code. The gap between these two sources concentrating on the application of the finite element method is now filled by several text books. Both the 'NAFEMS Finite Element Primer' [6] and 'NAFEMS Guidelines to Finite Element Practice' [7] have made a significant contribution. More recently, J.M. Steele's book 'Applied Finite Element Modelling' [8] has provided an excellent addition to the vast array of finite element textbooks already available. Brauer's compilation 'What Every Engineer Should Know about Finite Element Analysis' [9] addresses the wide range of applications of the finite element method.

THE ROLE OF PDA ENGINEERING

A course in finite element modelling must necessarily involve the use and understanding of a particular software product. Naturally enough we choose to use our own. This is a matter of convenience, rather than of our desire to promote our own products and we believe that this should not deter the attendance of users of alternative finite element analysis software. We feel that we are in an excellent position to provide a training course equivalent of the later text books mentioned above and contribute towards filling the gap which we see in the market of finite element training courses.

PATRAN Plus is an open ended, general purpose, 3-D Mechanical Computer Aided Engineering software package which offers one of the industry's leading finite element pre- and post-processors.

P/FEA is a multi-purpose structural analysis software package used for the solution of linear and non-linear static, dynamic normal mode, transient dynamic, frequency response, shock spectrum, steady state heat transfer, and buckling analysis of isotropic, orthotropic, or laminated composite material models.

Although we gear our course in finite element modelling towards the structural analyst, much of the material contained is relevant to anyone involved in finite element modelling. Exercises similar to those used to demonstrate features of modelling for structural analysis can be devised without too much difficulty for engineers who may be working in the fields of fluid or thermal analysis for example. To this end we also have available for use during the course, products such as P/FLOTRAN and P/THERMAL.

Although many people attending the course will have been users of PATRAN for many years, some may be completely new not only to PATRAN but to finite element analysis in general. Since PATRAN provides a pre- and post-processing facility to many analysis codes around the world, it does not require particular knowledge of any one system. Similarly, since P/FEA, in common with all PDA analysis modules, has a completely transparent interface with PATRAN, there is no need to get involved with input deck formats, etc. This means that it is possible to perform simple experiments on the effect of element shape, type etc. without a detailed understanding of the analysis code.

This combination of software allows increased confidence with a finite element software system to be achieved in a relatively straight forward way without distracting too much from the more important goals relating to general finite element modelling practice.

Within a large company there will undoubtedly be people with the experience to make decision on the building of a particular finite element model and who can try and impart this experience to others. However as technology changes, as new features are added to software such as PATRAN or any of the commercial analysis codes, these people may not have the opportunity to assimilate this information. As the use of finite element analysis continues to grow, there are are an increasing number of users who not only have no direct experience of modelling themselves but also have no immediate access to anyone who does.

All these issues point to the importance of training courses such as that outlined here. Of equal importance to the tuition received is the opportunity to take time out from the day to day pressure of producing results, reports, etc. and to stop and think about the issues we have discussed. Of course it is not possible to substitute many years of experience with a few days training, but it is possible to give sufficient knowledge that the student knows the right questions to ask in order to build up that experience as quickly as possible.

PDA have for many years been offering training courses in the use of the PATRAN Plus Software System which inevitably contain a high proportion of information concentrating on finite element analysis in general. More recently however we have come to the realisation that we should expand these aspects of our courses and offer training in finite element modelling which is not specific to any commercial software package. The remainder of this paper concentrates on what we believe the aims, objectives and content of such a course should be.

COURSE GOALS

The best way to summarise the goals in a course of this nature is to pose a series of questions which an engineer might ask before undertaking a finite element analysis.

- Is an analysis of this problem necessary ?
- Is there any way I can simplify the model ?
- How much detail should I model ?
- What type of element should I use ?
- What order of element should I use ?
- What type of analysis should I perform ?
- How should I distribute the elements ?
- What element density do I require ?
- What is the best way of transitioning from a coarse to a fine mesh ?
- How do I quantify cost versus number of elements ?

- How accurate are my results ?
- How do I best check my results ?
- What is the best way to display the results ?

Of course it is not the intention to answer all these questions here. However by describing some of the techniques which we use for our course, it is hoped that it will become clear how the students will be able to address them after four days of training.

THE ART OF FINITE ELEMENT MODELLING

In this section we shall examine some of the key points which we feel should be covered in a course of this nature. Many of the lessons are more easily taught by practice rather than by lecture. Indeed laboratory sessions should be an important part of any course of this nature. We will look in more detail at the format of these sessions later but let us first consider more generally the material that needs to be covered.

General Awareness

Time needs to be spent on discusing the nature of modelling and analysis in general. What is it we are doing when we build a numerical model of a real physical problem ? Since most people attending a course of this nature will be new to finite element modelling, it is a reasonable assumption that they will have little experience of numerical modelling of any sort. Hence we need to spend time talking about alternative numerical modelling techniques. What are finite difference methods or boundary element methods ? How do they differ from the finite element method ? How do I decide which method is best for my problem ? It is not necessary to go into great mathematical detail to explain these methods but it is important to put the finite element method into the context of the wider world of numerical analysis.

The key to success in all aspects of teaching is to bring your subject alive. If you have students who are genuinely wanting to learn you are a long way towards actually imparting some knowledge. A good way to achieve this in a course of this nature is to take some time out to explain a bit about the history of the subject and the areas where current research and development is concentrated. Pictures of real problems which have been solved using the methods are very useful especially if they reflect as wide a range of applications a possible. Any first hand anecdotal stories which the tutor can share are also a good way of setting up and justifying the content of the course.

We also take time in our course to explain the way that finite element analysis fits into the overall design cycle, with computer aided design and with manufacturing and testing (figure 1). It is also essential to get the importance of the analysis phase of the design cycle into context, to outline what it will do and perhaps more importantly what it will not do!

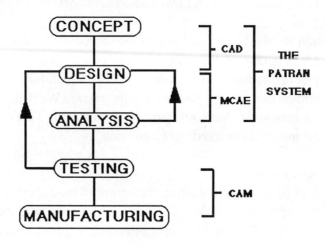

Figure 1. The Engineering Design Cycle

Defining the Problem

The next skill we need to consider is that of defining the problem that we wish to solve. All too often the first step in a finite element analysis of a component is to sit down at a computer and start to build a model. The importance of spending time up front to stand back and define; a) the actual physical problem and b) the problem which it is intended to model can not be emphasised too highly.

Simplification of the geometry is one of the most common steps to take but how do you teach this difficult task? Simplifications through the use of symmetry or approximating to a two-dimensional or axisymmetric geometric model are not too difficult. When it comes to actually modifying the geometry it is not so easy. Removing fillets, holes, and other detail are common practices in building finite element models but it is virtually impossible to give hard and fast rules on how this should be done. This is one area of the course where the use of examples can greatly help. Case histories discussed with the whole class and worked exercises can greatly help in putting across these ideas. It is not difficult to devise case studies where for example removal of a fillet can either simplify or over simplify the problem.

The modelling of boundary conditions is another area where great benefit can result from the presentation of case histories and the solution of suitable exercises. If we cannot represent the loading conditions in the model then it will be difficult to build and interpret the results of any finite element analysis at all. Most analysis codes have a limited set of boundary conditions which can be applied and from which the analyst must choose. It is no simple task to take these and generate a set of conditions for a problem which produce the same loading characteristics as those experienced in reality.

Creating a Suitable Finite Element Mesh

In an academic research environment, the ability to build the 'best' possible finite element mesh first time around is perhaps not the most important aspect of finite element modelling. Indeed experimenting with mesh density and distribution, element type etc. may be an important part of the research. In industry time seldom permits the luxury of being able to conduct these types of experiment - we need to get it right first time. The problem here is that the ability to build a good mesh is proportional to the experience accumulated through building and assessing finite element models of various types. If no time is allowed to experiment with different modelling techniques, then it is unlikely that a model will be the 'best' that can be built.

Of course as well as the general skills outlined above there are more specific ones such as transitioning the mesh density between different parts of a model and understanding how the behaviour of quadratic and linear elements or quadrilateral and triangular elements differ. Once again by far the best way to get this information across is by actually building some simple models and getting the students to try for themselves.

Materials Modelling

The most often voiced concern by engineers attending courses at PDA Engineering is the quality of the materials data they are using. Although software systems such as M/VISION [10] are now available to help with the storage, retrieval and selection of material data, the onus is still on the analyst to accurately build and use the material models. Since this could form a course in its own right, our finite element modelling course does not attempt to go into detail of the modelling of materials (especially composites) but merely to outline the issues involved.

Model Verification and Optimisation

We spoke earlier about the importance of careful planning before beginning to model a physical situation. Model verification is the time at the end of the modelling process where it is equally important to stand back again. We need to critically assess whether the model does represent the physical situation and whether the quality of the finite element model will be adequate for the intended analysis Guidance on the later is provided by trying to answer questions like 'what aspect ratio is acceptable for this element'. The PATRAN software has built in verification checks which make it ideal for this process (see appendix). We actually perform exercises where we take a model and change only the skew of the elements for example and compare results. Using these methods it is a relatively quick process to build tables of element performance for different element types, loading conditions and values of shape parameters such as skew or aspect ratio. For many users these tables can provide essential information as they head back to their place of work where they have to make decisions on mesh acceptability in isolation.

Optimisation of meshes for the purpose of reducing computer solution time is a good example of where it is difficult to avoid some theoretical work in order to explain the terminology and what it is we are actually doing when we renumber nodes or elements. Although I have continually put the emphasis on modelling aspects of this course, it is inevitable that some theory needs to be taught. This is discussed in more detail later.

Post-Processing

One of the most important parts of the whole finite element modelling process comes after the results have been obtained. Post-processing not only includes the presentation of results but also further verification of the model. Through careful post-processing of the results of an analysis it is possible to get much information about the quality of the finite element model and to decide whether or not the results are sensible. Do the results match intuition or any simple hand calculations if available ? Have the boundary conditions been applied correctly ? The user should use all the post-processing tools available to interrogate the model and answer questions of this nature. An important part of this process is to understand the errors that might be expected and to be able to assess whether or not they are significant. Once again a limited amount of theoretical background is required here. In our course we hold discussions relating to specific models and analyses and assess any errors there might be. Did we use enough elements? Were they the right sort of element? On observing the behaviour of the component could we have made any further simplifications? Was the solution time acceptable? The questions can go on and on and of course there is not usually any one correct answer. At this point we come full circle and are back to the realisation that finite element modelling is indeed an art.

Theory

So where does all this leave the traditional theoretical aspects of a course in finite elements. Undoubtedly there is still a very important role for these courses. It would certainly be a disaster if all finite element analysis became like a 'black box'. As we have already mentioned it is difficult for a single course to do justice both to the theory of finite elements and to the art of finite element modelling. Hence on a course such as that we have discussed here it is important to get the balance between the modelling and theory just right. This means enough theory so that all the modelling strategies can be understood but not so much that complex matrix algebra and actual element formulation for example, need be entered into.

THE ROLE OF EXPERIMENTATION AND EXERCISES

As will be quite clear by now, it is the use of practical laboratory sessions within a course of this nature that provide the vehicle by which many of the modelling techniques and strategies may be learned. Because we aim a course of this nature at anyone at all involved in the use of finite elements we need to be careful in using purely structural analysis examples. Ideally the students should be free to perform exercises which relate to whichever discipline they come from. This could be structures, fluid dynamics, thermal analysis, electrostatics, etc. The extent to which this occurs will vary from course to course depending on the experience and interests of the students attending. The ability to be flexible and have a variety of analysis software available is very important.

On a course of this nature it is essential to make full use of every opportunity to discuss any problem that arises as a result of performing the exercises. This may lead to the students devising their own tests to evaluate an idea they may have. Within the time restrictions of the course, diversions and experiments of this kind usually prove to be of benefit to all. It has even be known for the tutor to learn a thing or two on occasions !

The appendix contains an example of a typical exercise which might be given to students on a course of this nature. It deals with element verification and could well form the basis of an investigation by the student which goes well beyond the scope of the exercise itself.

THE FUTURE OF FINITE ELEMENT EDUCATION

Present trends in the use of finite element methods show that analysis is being performed increasingly earlier in the design cycle. Computer Aided Design software systems are starting to incorporate a limited analysis capability. This allows the designer to eliminate some of the preliminary iterations in the design cycle, giving him more confidence that the design which forms the basis of the first prototype is less likely to have to go back to the drawing office for modification. This popularisation of finite element methods means that an increasing number of people with no finite element background at all are building and analysing finite element models.

A course in finite element modelling would seem ideally suited to these people. Indeed they are unlikely to have the necessary skills to assimilate a theoretically based course. It is only through proper training in the use of finite element methods within the design cycle, especially in the area of modelling, that the most cost-effective use will be made of the analysis features within these design systems.

CONCLUDING REMARKS

I would like to conclude this paper by referring to Spooner's article in the October 1988 edition of Benchmark [11]. The purpose of that paper was to establish a syllabus for a course on understanding finite elements. As the author points out, in order to complete his syllabus (which includes a good deal of theoretical background) within a 30 hour timescale, it is necessary to keep tutorials and laboratory classes during the course to a minimum. It is recommended that exercises should be set but that they be carried out after the course. I believe that practice must assume the same importance as theory. To leave that practice to unsupervised work after a course invites the danger that the student will be diverted by other pressures. In addition, the benefit of interaction with the tutor is lost. It is for these reasons that I feel that an additional course along the lines of that outlined in this paper is desirable. From the stand-point of a company selling software for the pre- and post-processing of finite element models, this is the best way of ensuring that the theory is consolidated and can actually be put into practice.

Further information regarding any of the courses and documents discussed in this paper can be obtained for PDA Engineering International Limited.

ACKNOWLEDGEMENTS

I wish to thank my colleagues at PDA Engineering International for their assistance in the preparation of this paper.

REFERENCES

1. Mobbs, S.D., Carter, G.C., An Introduction to the Finite Element Method, Department of Applied Mathematical Studies, University of Leeds.

2. PDA Engineering, The Art of Finite Element Analysis, P/N 2192076.

3. Zienkiewicz, O.C., Taylor, R.L., The Finite Element Method, Fourth Edition, McGraw Hill,1989.

4. Heubner, K.H., Thornton, E.A., The Finite Element Method for Engineers, Wiley-Interscience, 1982.

5. Irons, B., Ahmad, S., Techniques of Finite Elements, Ellis Horwood, 1986.

6. NAFEMS, A Finite Element Primer, Department of Trade and Industry, National Engineering Laboratory, Glasgow, 1986.

7. NAFEMS, Guidelines to Finite Element Practice, Department of Trade and Industry,National Engineering Laboratory, Glasgow, 1984.

8. Steele, J.M., Applied Finite Element Modelling - Practical Problem Solving for Engineers, Marcel Dekker, 1989.

9. Ed. Brauer, J.R., What Every Engineer Should Know About Finite Element Analysis, Marcel Dekker, 1988.

10. PDA Engineering, M/VISION User Manual, P/N 2190011.

11. Spooner, J.B., Recommended Course for Understanding Finite Elements, Benchmark, October 1988, pp. 9-11.

APPENDIX

MODEL VERIFICATION EXERCISE

Introduction

Element verification is an essential tool for the analyst to check for and decide if the distortion of his model's elements will adversely affect the models performance to predict the effects of applied boundary conditions. PATRAN contains an extensive set of plate and solid element checks. It requires the user to assign threshold values for the checking algorithms so that the verification results will be specifically tailored to the users application.

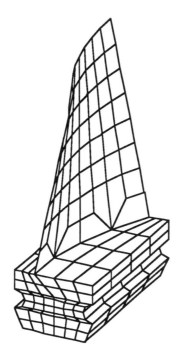

Figure 2. Geometry for Verification Exercise

Goal: To become familiar with various aspects of model verification and to determine the acceptability of the elements in the turbine blade model.

Exercise Procedure:

Step 1: Get into PATRAN and recall the turbine blade model you made in the earlier exercise

Step 2: Either by picking element verification from the **FEG** menu, or by typing the unprompted request **VER,** enter the verification menu. You'll be verifying elements so pick elements from this menu. Next pick verification by shape since you will be checking for elements with geometric abnormalities. (Special checks constitute Jacobian checks and higher order element checks.) Your model is defined with **QUAD** and **HEX** elements so you will want to verify both types of elements. Select **QUAD** to begin.

Step 3: Before you can perform the various element verifications you must specify the threshold values for the checks you will perform. You are required to manually set these values in PATRAN because only you know what deformities the elements can sustain and still produce correct results for your particular application. For this exercise use the following threshold values:

ASPECT RATIO = 3.5
SKEW = 20 (degrees)
TAPER = 0.5
WARP = 30 (degrees)
FSKEW = 20 (degrees)
EDGE ANGLE = 65 (degrees)

Step 4: It is good practice to record the results of you verification checks so that you can document the validation of your analysis or have quantitative values to use to iterate on a mesh that will give you good analysis results. Use PATRAN's text recording capabilities by setting print on. With print set on the verification results will be sent to the print file (**PATRAN.PRT**) as well as the alpha window.

Step 5: Check the aspect ratio of your elements. Answer yes when PATRAN asks if you would like to plot with the threshold set at 3.5. Next perform a **FILL HIDE** plot. Since you have told PATRAN to plot with using a threshold value for this test, any elements PATRAN finds with an aspect ratio greater than 3.5 will be colour coded red to indicate that those elements did not pass this test.

Step 6: Exit from the plotting menu and choose the skew check. Repeat the steps described in Step 5.

Step 7: Exit from the plotting menu and choose the taper check. Repeat the steps described in Step 5

Step 8: Exit from the plotting menu and choose the warp check. Repeat the steps described in Step 5

Step 9: Exit from the plotting menu and the QUAD menu and start checking the HEX elements. Choose the face skewness check. Repeat the steps described in Step 5

Step 10: Exit from the plotting menu and perform the edge angle check. Repeat the steps described in Step 5.

Step 11: Now that you have finished verifying the model's elements remember to set PRINT OFF to stop sending information to the print file.

Step 12: To see your applied temperature distribution clearly perform a colour assignment plot of the temperatures with the load set ID you defined in the earlier exercise. Once you've assigned colours to the temperatures perform a FILL HIDE (FH) plot of your model.

Step 13: To see your applied pressure distribution clearly perform a colour assignment plot of the pressures with the load set ID you defined in the earlier exercise. Once you've assigned colours to the pressures perform a FILL HIDE (FH) plot of your model.

Step 14: To see the thickness distribution clearly perform a colour assignment plot of property record 2. Once you've assigned colours to the thicknesses perform a FILL HIDE (FH) plot of your model.

Step 15: Either by picking optimisation from the **FEG** menu, or by typing the unprompted request **OPT,** enter the optimisation menu. You want to optimise for ABAQUS so pick the appropriate optimisation schemes from the ACTION, METHOD and CRITERION menus:

UK HEI STRUCTURAL MECHANICS FINITE ELEMENT USAGE AND NEEDS

RG ANDERSON and JW MARTIN
Manchester Computing Centre
University of Manchester
Oxford Road, Manchester M13 9PL

ABSTRACT

The results of a survey of the usage and needs of structural mechanics finite element codes in Higher Educational Institutions (HEIs) are reported and discussed. 50% of HEI users choose either the UK code PAFEC or the US code ABAQUS. PAFEC is popular for teaching and ABAQUS for research. Some researchers need more support than the Academic license provides. The terms of the Academic license may be re-negotiable to cover support and to clarify US export controls on open academic work. A proposal to set up an HEI Community Club to help users to focus their needs to developers is made.

INTRODUCTION

Computational Structural Mechanics (CSM) plays an important and increasing role in the design and assessment of engineering structures by UK industry and HEIs. Industry employs several thousand engineers at a cost of order £100m/annum committed to using such methods at various levels of sophistication. The firms range in size from very large, hi-tech industries each with over a hundred engineers using CSM (eg GEC, BAe) to consultancies with one or two people.

CSM Finite Element (CSM FE) codes were started in small research groups in HEIs in the US, UK and Germany in the early 1960s. Commercial CSM FE codes are now developed in medium sized organisations often with 100 or more engineers engaged on development and support together with a world-wide spread of selling agents. In 1990, one of the major US developers, MSC-NASTRAN, reported revenues of $45m in 1990.

Although commercial code development (eg ABAQUS, PAFEC) is now rarely carried out in HEIs, usage of commercial codes for research and teaching in HEIs is about one-half of the industry

160

usage. Information from Pollard [1,2] of the Queen
Mary/Westfield College Centre for Engineering suggests that in
early 1991 there were perhaps 800 Academic staff and 800
Research Students using CSM to research structural behaviour in
Universities. The Polytechnics contribute about another 25% to
these University numbers to make up the full HEI community..

The most frequently cited research interests in the
community are pressure vessels, automotive engineering,
offshore applications and metal forming. In addition there is a
growing teaching load. The data of Pollard [1] suggested that
in 1989 there were about five thousand students being taught
finite element topics at various levels from 1st year to Taught
Postgraduate. By early 1991 a second survey (Pollard, [2])
suggested these numbers had increased to between six and seven
thousand students. The main engineering disciplines teaching
CSM FE are shown in Figure 1. The data from Pollard [2] suggest
that the teaching and research effort in Universities was
backed by about 1000 dedicated work stations or PCs plus
mainframe and supercomputer access.

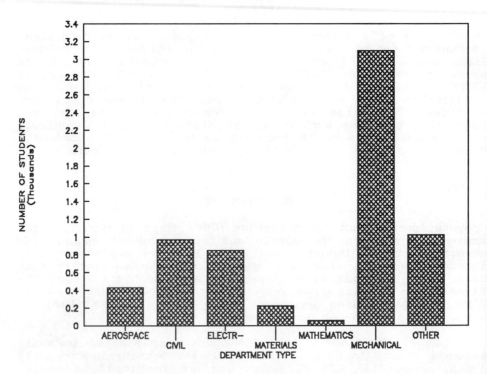

Figure 1. The distribution of University students by Department
(with acknowledgements to D Pollard [2])

The predominance of mechanical and civil engineers is not unexpected. The large number of students (comparable to Civils) receiving teaching in Electrical Engineering finite elements will be studying electrical codes as well as CSM FE codes.

The output of graduate and post-graduate engineers with training in CSM from HEIs has the potential to double the numbers available to industry over the next few years. The HEI usage is therefore very important to industry, to code developers (who wish to see future industrial users choosing their codes) and to hardware suppliers.

THE IMPORTANCE OF US COMMERCIAL CODES AND THEIR PROBLEMS

The survey results (see next Section) suggest that over 50% of the UK CSM FE community choose US commercial codes. Whilst all users of commercial codes are dependent on the developer to provide his exact needs, users of US codes in the UK have also to consider the impact of US export controls.

In particular, when using US origin goods and technology, the UK user must consider the re-export requirements of the US Export Administration Regulations (EARs). These Regulations control the re-export of US goods and technology (including software) and their "direct product". The Regulations should be seen as one facet of international control of hi-tech products which began with the COCOM controls directed against former Warsaw Pact countries and which are now being actively extended to prevent proliferation of military information to sensitive Third World countries. The UK shares the concerns of the US on proliferation and, of course, exercises control over exports from the UK. The UK continues to object in principle, however, to extraterritorial US export controls, while generally allowing UK individuals and companies to make a commercial decision on whether to comply with US licensing requirements. Some HEI users have interpreted the "direct product" control of the US EARs as limiting the open publication of research using the US codes.

The application of the US EARs, which is exercised within the license conditions, varies considerably from one developer to another. There is some indication that the Academic licenses giving the user least scope for conducting research are subject to fewer extraterritorial controls. One license makes no reference to the EARs. It is concluded that relief should be obtainable from the re-export controls in the US EARs when the Academic license meets certain criteria. It is important to establish these criteria with both the US Department of Commerce and the software suppliers to ensure acceptance.

THE SURVEY

The Survey was conducted in the period May to September 1990 using three distributions (SERC Engineering Applications Software Environment contacts, Inter-University Software Committee Computer contacts and the NAFEMS BENCHMark UK distribution) which did not favour particular codes.

There were 157 HEI responses (c.8% of the HEI community) and 34 responses from industry (c.1% of industrial users). The HEI sample is excellent (the sample of University staff is about 15%) and the industrial sample is acceptable for drawing broad conclusions. The responses came from 79 University Engineering Departments out of 186 Departments identified by Pollard [1] and the present Survey as possible users of CSM FE codes. 13 of the replies came from 12 Polytechnics (which were not covered by Pollard's surveys [1,2]).

Distribution of replies amongst Departments
The distribution of survey replies between Departmental disciplines is shown in Figure 2.

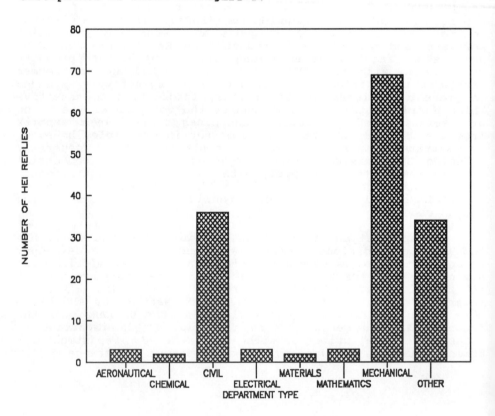

Figure 2. The distribution of HEI replies by Department

Distribution of replies amongst staff grades

The distribution of staff grades is shown in Table 1 and is fairly typical of a University Department staff. There were no undergraduate replies.

TABLE 1
The staff grade structure of HEI replies

Academic Grade	Replies (%)
Professors	9%
Readers/Senior Lecturers	21%
Lecturers	32%
Post-doctoral researchers	19%
Research students	19%
Undergraduates	nil

Code choices

Country choice: The country of origin of the most used codes is of considerable interest to policy-makers. No codes from mainland Europe or the Far East were cited as first choices and only three users cited them as "available" out of 304 such citations. The code usage is, instead, split between US and UK codes as shown in Table 2.

TABLE 2
The usage of US and UK codes in HEIs and industry

Type of code	Country	HEI	Industry
Commercial	US	45%	75%
Commercial	UK	37%	25%
In-house	UK	18%	small

The industry sample is sufficient to attach weight to the choice between US and UK origins. 75% of the industrial market cited US codes whilst the HEI users of commercial codes chose US codes over 50% of the time. In-house codes attract about 18% of HEI users both for research into specialised fields and for teaching.

Code choice and availability: The industrial sample is too
small to distinguish between the popularity of five leading
codes (ANSYS, PAFEC, ABAQUS, NASTRAN, NISA). The HEI users'
choice is much clearer and is shown in Figure 3. Users were
asked to cite their code of first CHOICE and other codes
AVAILABLE to them. 156 first choices were cited and 304 codes
were available to the 157 HEI users (ie two codes for the
average user).

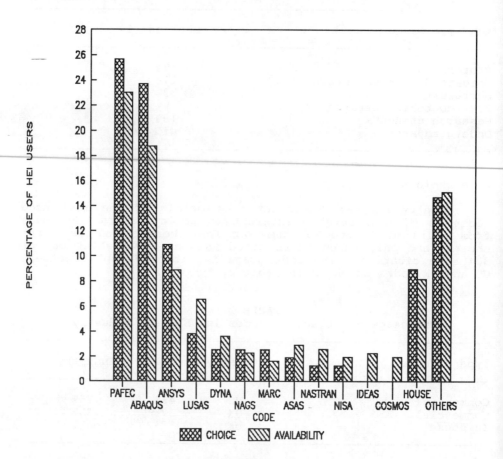

Figure 3. The twelve leading codes used in HEIs

Comparison of PAFEC and ABAQUS
The two codes (PAFEC (UK) and ABAQUS (US)) took 50% of the HEI
market between them. Follow-up discussions at Nottingham
University and at UMIST suggested that the codes are
complementary rather than in direct competition for the same
market. The key features of interest to users were as follows:

1.PAFEC is extensively used for design associated with quasi-linear problems, for teaching and for project work.

2.ABAQUS is used extensively for non-linear research activities. Circumstantial evidence built up during the follow-up confirmed the importance of ABAQUS as a non-linear code.

Computer choices
The choice of computer manufacturer extends from super-computers to PCs. The largest fraction of citations (25%) were for VAX machines (many of them small mainframes). APOLLO, with 15% of citations, was the largest supplier of work stations. The Manchester Computing Centre National facility - the AMDAHL VP1200 - was cited by 12.5% of users (mostly ABAQUS users). Pollard [2] has provided raw data from a hardware survey conducted in early 1991 which suggests that large purchases of work stations and PCs are currently being made in Universities.

Pre/post processor choice
About 65% of HEI users employed a commercial pre/post-processor which gives them an interactive facility to set up meshes and to look at results. The remainder use batch methods. Some of the pre and post-processors are specific to one code, others use neutral files to couple into the major codes.

The survey showed that PAFEC is usually coupled to the PAFEC pre/post-processor PIGS. The most cited general purpose codes were PATRAN, IDEAS and FEMVIEW. About 50% of ABAQUS users cite PATRAN as an interactive pre-processor.

Computer usage
There have been many surveys of codes installed on computers. This survey attempted to characterise the use made of the codes. The features of usage reported in Table 3 are:

1. Number of runs made per week
2. Average length of a run (in hours)
3. Average number of elements used in a problem
4. Average number of degrees of freedom used in a problem

TABLE 3
Computer usage

Feature of usage	Industry All users	HEI All	HEI PAFEC	HEI ABAQUS
Number of replies	34	157	40	37
Runs/week	33	15	23	9
Averages				
run length (hours)	2.6	2.1	1.3	1.8
elements/problem	1070	585	760	440
degrees of freedom	7972	5272	5770	6602

Table 3 requires some additional comments:
1. A number of the runs by PAFEC and ANSYS users were for teaching or project purposes. These teaching/project runs are not properly shown in Table 3 and the Table should be read as applying to research activities. A crude estimate suggested that the average number of research runs using PAFEC should be reduced to 16/week/user and the average run-time increased to 2 hours.

2. The "average" computer cited in the survey was a slowish work station (both for HEIs and industry). More than half of the ABAQUS runs were done on the AMDAHL VP1200 which is very many times faster than the "average" computer cited in the survey. The average ABAQUS run time (1.3 hours) thus represents a lot more computing than the average HEI user run time of 2.1 hours. Since the average ABAQUS problem size in Table 3 is comparable to the HEI average, it must be hypothesised that the average ABAQUS user is solving much more difficult (ie non-linear problems) than the average HEI CSM FE code user. This hypothesis will be further tested under "non-linearities".

3. A second measure of complexity is whether the problem being solved is three-dimensional. Most users said they were solving problems in three-dimensions. Since the average HEI user has 9 degrees of freedom/element it seems that many problems must be two dimensional. The average ABAQUS user with 15 degrees of freedom/element is more likely to be solving three-dimensional problems than the average HEI user.

4. Using the values in Table 3, it is possible to do a very simple calculation to obtain the number of computers being used for CSM FE University research. The average user takes up 30 hours/week. The total usage for a sample of 15% (remembering that 13 replies came from Polytechnics) is then

Number of computers in full time use = (30/168)*144/0.15 = 171

Since there are about 180 engineering departments this calculation suggests that there is an average of 1 work station per department being used full time for CSM FE research.

Materials modelled
50% of users cited steel (albeit of many different types). Aluminium (14%), composites (14%) and concrete (10%) dominate other usage with plastics, rubbers, semi-conductor materials and ceramics all receiving a few mentions. One user cited "bone". The list of materials included many requiring non-linear descriptions and a good code will have a facility to enable the user to write in-house non-linear material rules.

Non-linear facilities
Non-linear facilities are becoming increasingly important for structural assessment purposes. The main non-linearities are geometrical (eg large strains and deformations), material (eg plasticity, creep) and boundary discontinuities (gaps, friction, cracks).The number of citations of non-linear

applications in HEI research is given in Table 4 for the four leading HEI codes (ABAQUS, PAFEC, ANSYS and LUSAS).

TABLE 4
Number of citations of non-linear applications

Application	ABAQUS Number	PAFEC of	ANSYS citations	LUSAS
Non-linear geometry				
Large displacements	26	15	7	4
Large strains	20	8	3	2
Large rotations	15	2	3	1
Post buckling	10	3	5	1
Non-linear materials				
Elastic-plastic	26	13	7	3
Elastic-visco-plastic	9	1	2	0
Creep	6	3	4	1
Composites	15	1	6	0
Boundary non-linearities				
Contact problems	17	9	4	2
Friction	8	2	3	0
Total non-linear citations	126	42	37	10
Total respondents for code	37	40	15	6
Total citations/respondent	3.4	1.1	2.5	1.7

A strong association between ABAQUS users and non-linear calculations can be seen in Table 4 with each user dealing with a multiplicity of non-linearities.

Use of CSM FE codes in teaching
The survey dealt primarily with uses of CSM codes in research. The extent of teaching of FE had been dealt with by Pollard [1] in a survey of departmental usage (rather than individual usage). Their use in teaching can be identified in about 10% of the replies.

The key problem facing a teacher is whether to teach the principles using in-house codes or whether to teach a commercial code with a vocational objective. Both approaches are well taught in a number of Departments. In view of the expanding requirements for trained CSM FE users, it is important to build on the experience of the successful groups.

Code problems
Code problems were assessed in two stages. The users were first

asked whether they had any of the typical (but not exclusive) problems in running the current versions of their codes. They were then asked to list desirable code features which would lead to them making greater use of CSM FE codes.

Typical problems: The response to a list of typical suggested complaints is shown in Figure 4:

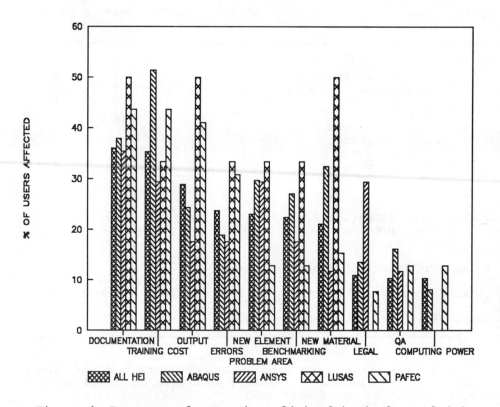

Figure 4. Response of users to a list of typical complaints

The wording of the suggested complaints in Figure 4 was:

Typical support complaints
    The code documentation is inadequate
    Training is too expensive
    Post-processing facilities are inadequate for design
    Benchmarking near the user's problem is unavailable
    There are legal problems in using the code
    Quality Assurance is inadequate for the user's problem
    Code unavailable to the user on a supercomputer

Typical functionality complaints
    There is no error estimation facility
    New elements are awaited
    New material models are awaited

Most complaints were directed to code Support with fewer complaints about Functionality. There was little difference between the responses of the users of the four leading HEI codes. These problems were, of course, leading questions. Users were also given the chance to state their own problems and future needs.

The following list of frequently cited additional problems was compiled from the HEI response:

Coupling between codes (eg fluids/structures, pre/post) is poor
Lack of access to some areas of source limits use
The diagnostics are poor
Costs of general pre/post processors and documentation are high

Follow-up
The Survey was followed up by visits to two groups making extensive use of PAFEC (Nottingham University) and ABAQUS (UMIST). As a result of these visits it became clear that a number of the Support complaints were symptoms of the limitations of the Academic license offered at cheap rates by developers. In particular, the Academic license provided insufficient cover for some members of the UMIST group who were undertaking very non-linear problems using ABAQUS.

It is suggested that an option to obtain additional support from a code developer should be made available to users taking the most advanced codes to their limits for research. This should include options for additional training facilities. This would involve those taking up the option to pay more than the basic Academic license fee (but much less than the commercial license fee).It has already been suggested that some Academic licenses need reconsideration because of US Export controls and there is a strong case to negotiate such deals centrally. It would seem unlikely that codes used only for teaching would need additional support.

USERS' FUTURE NEEDS
The users were asked to state which features would lead to more use being made of CSM FE codes in the future. These needs have been brought together with the results of follow-up visits, the recent NAFEMS Survey on Research Needs and informal discussions with academics, industrialists and developers to draft an agenda for the sort of CSM FE code that will be needed for the year 2000. It should be emphasised that this is an agenda to assist developers to spend money wisely rather than a proposal for the academic community to spend a lot of money. The way in which the Agenda is to be handled will be dealt with at the end of this section. The background and agenda for "FE2000" is:

"FE2000" will be used by a community two or three times the present size. Two areas where activity is expected to grow rapidly are Fast Design and Assessment:

1. Fast Design is needed by manufacturing industry to lead world competition. It will be conducted within a highly integrated CAD/CAM environment with coupled heat transfer, CFD, electromagnetic and other physical inputs. Automatic meshing/remeshing and optimisation will be standard. Output will cover UK/EEC Standards, CAM and, no doubt, draft the final design report

2. Assessment using extensive non-linear functionality is increasingly needed for manufacturing feasibility, structural integrity, operational life assessment, post-failure and other analyses. Post-processors will relate to Safety cases and other output needed by designers, manufacturers and Regulatory Bodies.

The user needs for the future can be listed as follows:

Pre/post processing:
The interactive solid modelling capabilities of the CAD/CAM majors
Automated, refined, specialised (eq crack) meshing
Interactive post-processors up to reporting of results
Graph plotting and multi-media capabilities eg video

General:
Compatibiity with the complete design process (modeller to report)
Error estimation and targetted adaptive remeshing
Optimisation trends
Large problem capability (Substructuring, supercomputer versions)

Non-linear requirements:
Geometry
Materials (associated with material testing procedures)
Boundaries eg Gaps, cracks, slide-lines
Radiation (with reflection)
Large problem capability

Code features:
User defined materials, elements
Transient, cyclic and stochastic history inputs
"Shake-down" methodology
Access to some areas of source code
Easy interfacing to comparable codes for verification
Economic features eg shell elements
Wave front optimisers (retained for a given problem)

Code documentation:
User, Example, Verification, Theory, Programmers Manuals
Verification manual to include all recognised benchmarks
QA plan and procedures
Developer newsletter

Code support:
Bug-list
On-line HELP facility
"HOTLINE" telephone help
"Expert system" back-up to HELP and "HOT-LINE"
At least Annual releases with intermediate bug fixes.
Access to developer to obtain financed development work
Cheap elementary training
Advanced training
World and local user groups and meetings
More Benchmarking to meet wider community interests
Audit access to QA

Computers:
PCs, workstations, supercomputers, networks
Software for translating binary between computers

Costs:
Competitive licensing, support, training and running costs
HEI deals including site licensing, support and training

Code versions:
There is a need for versions ranging from undergraduate, project and industrial training to versions tuned to the fastest computers to obtain quick turn round.

Licensing:
Academic (and industrial licensing) should be subject to clear restrictions under COCOM, US and other international export controls.

Large groups of work stations on one site (particularly academic sites) will make site licensing imperative. This should be subject to the minimum of physical controls eg dongles.

Academic licensing which limits problem size allows teaching but can make research expensive This trend needs to be watched carefully.

Progress towards completing the agenda for "FE2000" is most likely to come from an amalgam of competitive efforts from several code developers, CAD/CAM modellers and hardware vendors coupled to constructive and timely feed-back from users. The survey has shown that the HEIs are currently playing an important seminal role in the UK CSM FE community with many of their activities parallel to those of industry (eg fast studentprojects, advanced analysis). It is therefore appropiate for the HEIs to play a constructive role in the development of, for example, fully integrated codes over the next few years. With the objective of assisting the whole community towards these developments, the following suggestion has been inserted into the full report made to the Computer Board:

172

PROPOSAL FOR A COMMUNITY CLUB

An HEI Community Club for CSM FE should be set up to feed back user response on the support, training and functionality of CSM FE codes and their integration into CAD/CAM over the next few years. The Club would include users from HEIs and industry, the SERC, developers across the CAD/CSMFE/CAM field and hardware vendors. The Club would complement NAFEMS efforts on research, standards and verification as well as those of code oriented User Groups, Special Interest Groups and other related groupings. The SERC run a CFD Community Club which, although serving a more academically oriented community is a useful prototype with extensive European connections.

ACKNOWLEDGEMENTS

The Survey funding was provided by the Computer Board.

The Survey organisers would like to thank

1. Drs P.R.Smith and D.E.Pollard of the Computers in Teaching Initiative Center for Engineering at Queen Mary and Westfield College, London for helping to build a picture of the HEI CSM FE community

2. NAFEMS and EASE for distributing the surveys

3. the respondents for replying to the surveys.

Mr C Cartledge, Chairman of the Inter-University Software Committee CAD/CAM Working Group has provided continual support throughout the survey and its follow-ups.

Professors TH Hyde (Nottingham), SR Reid (UMIST), B Parsons (QMC/Westfield) and Dr BJ Marsden (AEA Technology) organised follow-up visits and took part in helpful discussions.

Professors JT Boyle, CR Calladine,FRS and GAO Davies made constructive comments on drafts of the survey report.

The organisers have been advised on matters relating to the US Export Administration Regulations by N Cooper and M Grew of the North American Trade Policy Branch of the DoTI.

REFERENCES

1. Pollard, D.E., Smith, P.R. and Brandon, J.P., Computer usage in Engineering Education in UK Universities, publication form the CTI Centre for Engineering, Queen Mary and Westfield College, London, January 1990.

2. Pollard, D.E., Unpublished survey data made available to RG Anderson, January/February, 1991.

THE DESIGN AND ANALYSIS OF HELICAL MILLING CUTTERS USING INTEGRATED FEA PACKAGES

D.C. WEBB and K. KORMI
Industrial & Manufacturing Division,
Faculty of Information and Engineering Systems,
Leeds Polytechnic, Calverley Street,
Leeds LS1 3HE, UK.

ABSTRACT

We describe here the use of 3 integrated FEA packages to analyse the design of helical milling cutters. This particular problem was chosen as an example to demonstrate the use of FEA to students taking CAD and FEA modules on a Manufacturing Systems Engineering B.Eng. course.

The packages used are FINEL, ABAQUS and FEMVIEW/FEMGEN. The mesh is constructed using the macro command in FINEL. The model generated is then transferred to ABAQUS where appropriate loads and constraints are imposed and a stress analysis carried out. The results of the analysis (stresses, strains and displacements) is then graphically presented by using FEMVIEW/FEMGEN.

We discuss how, by changing various parameters in the input data, this method of analysis helps enable students to appreciate various aspects of FEA and its importance in the design process.

INTRODUCTION

The study of Finite Element Analysis (FEA) offers the opportunity for study in many different academic areas - applied physics, engineering mechanics, structural mechanics, numerical computation, etc. It is a standard, well known and widely used technique that is taught on many engineering related courses. The Finite Element Method (FEM) is, of course, already widely used in industry and increased public awareness of safety together with a desire for companies to save money on building prototypes and/or testing to destruction have made it particularly relevant to the contemporary design process.

However, a major problem with using the FEM is its complexity. The amount of theory and mathematical development required to give an in depth knowledge of the subject is daunting - especially to those who have little or no relevant knowledge or experience in mathematics and advanced numerical and computational techniques. At Leeds Polytechnic we are attempting to teach more of the FEM to students with less background knowledge of the mathematics required to fully understand the theory.

<u>Course Background:</u> The direction and content of engineering courses at Leeds Polytechnic is following the same route as those at many other establishments of higher education. Difficulty in recruiting students for engineering and the recent popular trend for business orientated courses has changed the name of the relevant engineering course within our faculty of "Engineering and Information Systems" from "Production Engineering" to "Manufacturing Systems Engineering". Next academic year we will be making further changes to a modular degree course that will offer three routes - "Manufacturing Systems Engineering", "Manufacturing Systems Engineering and Business" or "Manufacturing Systems Engineering and Automation".

The strategy of educating a future generation of engineers is a matter for discussion elsewhere and we comment on it here merely to attempt to present an idea of the type of background that we have come to expect of students to whom we are required to teach FEA.

As a consequence of these directional changes we find ourselves teaching more and more students with less and less of a rigorous scientific or mathematical background. We therefore find that in the final year of the course - where a FEA module is offered as an option - we are faced with students with varying levels of knowledge and understanding of the relevant disciplines. As a consequence of this, and because there are now a number of well tested and well documented computer packages available for FEA, we tend to use a "black box" approach which concentrates more on the sort of things you can do with FEA and interpreting results, than on how the technique works. Of course, it is also important that the technique is described - as there will be some students able to understand and appreciate the elegance of the mathematical methods used - but the main emphasis is on the use of FEA.

THE SOFTWARE PACKAGES

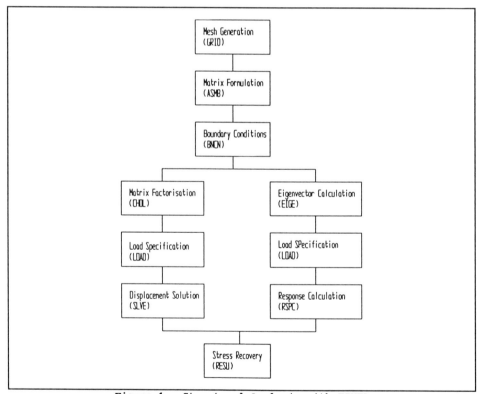

Figure 1 - Structural Analysis with FINEL

The FEA packages used at Leeds Polytechnic are FINEL and ABAQUS. FINEL was initially conceived and written in the Department of Aeronautics, Imperial College and was developed commercially by

Babcock Energy Limited (later to become Crawley and Renfrew). Although it is no longer supported it is easy to use and offers a good range of modules for both static and dynamic analyses. It also incorporates reasonably straightforward methods of constructing geometry. A finite element mesh is generated by dividing the object into simple shapes (taken from entries in a Region Library) and then dividing them into elements.

Figure 1 (taken from the Finel User Reference Manual [1]) shows the modules available in FINEL and illustrates how an analysis is conducted by using a sequence of them. Because of its ease of use FINEL is an extremely effective package for teaching the basics of FEA.

ABAQUS originated from the US in the mid 70s and has since become one of the most thoroughly tested and versatile FEA packages

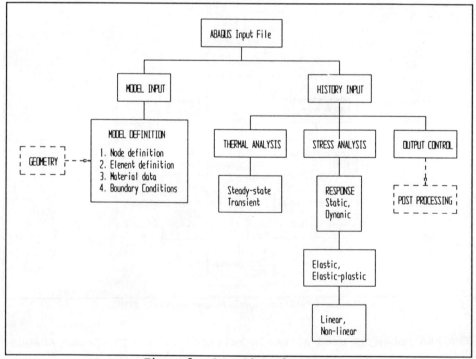

Figure 2 - Operation of ABAQUS

commercially available - with over 10,000 users world wide. It is an extremely powerful package and therefore more complicated to use than FINEL - requiring many hours of 'hands on' experience. ABAQUS is however, a very good example of the 'state of the art' in FEA packages - and, because it is more comprehensive, it can be used to demonstrate the power and flexibility of the FEM. The operation of ABAQUS is outlined in Figure 2 (adapted from a student project report [2]). The input file provides the ABAQUS package with information on the model and the history (or type of analysis to be performed). Data can be input to assist with model and/or load definition and can also be output for post processing.

In addition to these packages, we have pre and post processors available in the form of **FEMVIEW/FEMGEN** and **SUPERTAB**. The latter also interfaces with **GEOMOD** which is a full 3-D solid modeller from the **IDEAS** suite of packages.

THE FEA TEACHING MODULE

As with most courses of this type, the FEA module begins with a brief description of what the FEM is. A number of preliminary simple examples can be used with FINEL to illustrate the important points of the package (mesh generation, boundary conditions, loading, etc.). Major considerations at this stage are how to present the data to the computer package; how to use the various package modules and how to interpret the results that come from the analysis. In this way students can become familiar with some of the basic problems that FEA can solve.

In some of these early examples, theoretical solutions and experimental results can be compared with those from FEA. As more examples are introduced however, the complexity of the model can be increased and discussion directed in more detail towards the meaning, extent and reliability of the results from the computer rather than on performing experiments or going through complex theoretical equations. However it is important to emphasise the physics of what is happening in each case so that students are able to 'tell the story' behind each example.

One way of helping to achieve this by a series of workshops which augment the lectures and through which students (or groups of students) can work more or less at their own speed. These workshops build up eventually to the introduction of the ABAQUS package to perform a series analysis of a complex model. What follows is a description of one particular example FEA problem that students can identify with, formulate and solve.

THE HELICAL MILLING CUTTER

The design and analysis of helical milling cutters is of particular interest to those involved in a manufacturing environment. A great number of products that are manufactured involve components which, at some stage or other in their construction, are exposed to the milling process. Thousands of milling cutters exist and are made each year. An analysis of the stresses that are exerted upon these cutters can give information that can be utilised in their design and help us to predict their lifetimes etc. In particular, this example:

- illustrates how different packages can be integrated to help solve a complex problem - it also highlights the problems that exist with integrating packages that have been developed for different purposes;
- uses a component that students are familiar with;
- can be used to show how varying the geometry of the tool alters the degree of stress experienced;
- can be used to demonstrate how stresses and tool geometry react to changes in depth of cut, spindle speed, feed rate, etc. - all examples of situations that students will be familiar with;
- allows a demonstration of how, by carrying out such an analysis, FEA can be used as a design tool.

The Model
The fully developed model of the helical milling cutter is shown

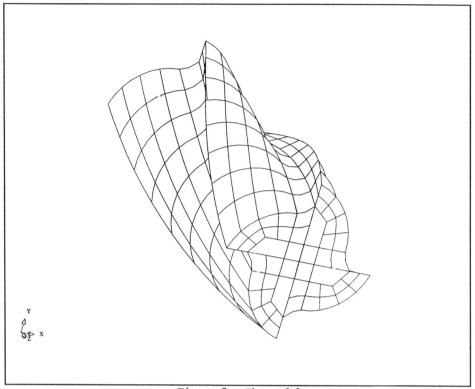

Figure 3 - The model

in Figure 3. Another reason for choosing this component as an example is that the generation of the geometry can be used to demonstrate the various approaches (and their relative advantages and disadvantages) to creating geometry that are available in the different computer software packages that we have at our disposal.

Creating the Geometry

Using GEOMOD/SUPERTAB: A 3-d solid model of the geometry can be created relatively easily using GEOMOD. First, a 2-d profile of one quarter of the cross section (shown in Figure 4) is created and this is copied to the other three quadrants (Figure 5) to give a profile of the complete cross section. The full 3-d model is then created by extrusion and rotation. The amount of extrusion and degree of rotation can be varied to produce a particular helix angle. In GEOMOD it is possible to paramaterise this process of constructing the model by using a 'program file'. The user can then be asked to enter parameters such as the extrusion distance and degree of rotation - and will therefore be able to generate a whole family of similarly shaped cutters.

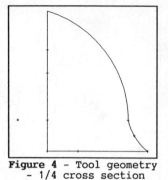

Figure 4 - Tool geometry - 1/4 cross section

A difficulty with this method of creating geometry for FEA does arise with the next

step - generating the mesh. This can be achieved by transferring the data to SUPERTAB which enables nodes and elements to be constructed around the geometry of the 3-d GEOMOD model. This information can then be written to a file for input to ABAQUS for analysis.

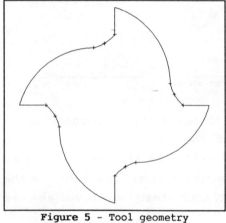

Figure 5 - Tool geometry cross section

This particular route is, however, perhaps not the best one. Although SUPERTAB can be used to generate a mesh, it is a large and fairly complicated package in its own right, and a certain amount of training and familiarisation is required before it can be used effectively. There is not enough time on our course to become familiar with all of these large complex packages. Therefore this method of geometry creation is better given as a demonstration as it is easier and quicker for students to use the methods of element generation that are linked more closely to the definition of component geometry - i.e. by using FEA packages themselves.

Using FINEL and ABAQUS: To create the model and analyse this problem, two different aspects of the two FEA packages are employed. The FINEL package is not powerful enough to perform the analysis and this must be carried out by ABAQUS. However, the geometry is quite complex and cannot easily be generated in ABAQUS although it can be readily created using the macro techniques available in FINEL.

In this case the geometry can be generated in three stages:
• the creation of a 1/4 2-D cross section tool geometry and corresponding nodes;
• transforming and copying these nodes to enable 3-D elements to be created to form a 1/4 3-D tool cross section;
• transforming and copying these 3-D elements to form the complete 3-D model.

Generation of the geometry in these stages requires the use of transformation macros in the FINEL package. Let us look more closely at each of these stages in turn.

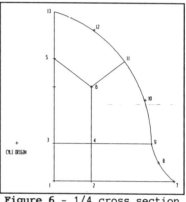

The first stage - the creation of a 1/4 tool cross section of is used to define the 2-D node positions. Figure 6 shows the situation.

Figure 6 - 1/4 cross section geometry creation

The nodes are numbered here in the order that they are created in the FINEL input file. A mixture of Cartesian and Polar coordinates are used to help define the geometry (see below) and the FINEL regional library entries SQUD and PQUD are used to generate the 2-D Plane Stress/Strain Elements of type PM08 (see Figure 7). SQUD is a straight sided quadrilateral, defined by 4 vertices. PQUD is a parabolic quadrilateral defined by 4 vertices, 4 edge mid-points and 1 face mid-point.

In the second stage, the 2-D cross section is transformed to generate enough nodes in 3-D to enable a quadrant of the tool to be defined in terms of 3-D HX20 elements. The amount of

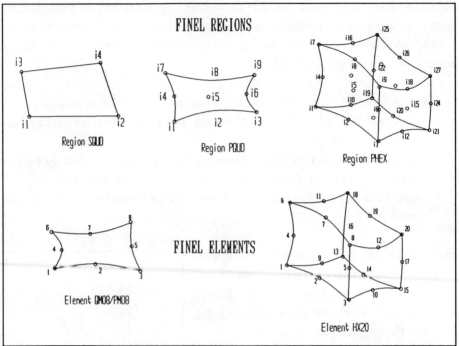

Figure 7 - Element types used in Finel for this analysis

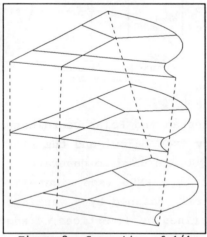

Figure 8 - Generation of 1/4
3D cross section

translation and rotation of the 1/4 cross section is defined by the helix angle, the resulting planes of nodes generated are shown in Figure 8 and the corresponding elements shown in Figure 9. The FINEL input file used to create the nodes in Figures 8 and 9 is shown below. In this case the 2-D Plane Stress/Strain Element PM08 has been replaced by the 3-D Membrane Element QM08 - so that the transformations in 3-D can be carried out. The basic 1/4 cross section geometry is defined within a macro named 'ONE'. Macros can be used in FINEL to facilitate the generation of meshes that are composed of repetitive sub-units. The translation and rotation are performed by using the macro commands TRANS and ROLL as shown.

Finally, the full 3-D geometry is generated by using the nodes defined previously and translating and revolving along the body of the tool to create the model shown in Figure 3. The node co-ordinates for the full 3-D model are thus generated by FINEL and can be transferred to an ABAQUS input file for further analysis by using the relevant FINEL command.

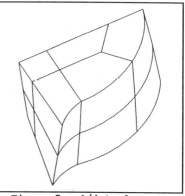

Figure 9 - 1/4 tool cross section

```
ANAL GRID
DATA BASE ZTL
ELEM PM08 1.0
REGI COOR
   0.00000      0.00000
   2.50000      0.00000
   0.00000      2.00000
   2.50000      2.00000
   0.00000      6.50000
   2.50000      5.00000
   6.50000      0.00000
   5.65000      1.00000
SYST POLA  -2.00000   2.00000
LET R0 7.3
REGI COOR
   R0       0.00000
   R0      18.50000
   R0      37.00000
   R0      55.50000
   R0      74.00000
SYST RESE
MACR STAR 'ONE'
MACR STAR 'TWO'
REGI SQUD  1  2  3  4                    1  1
REGI SQUD  3  4  5  6                    1  1
REGI PQUD  2  0  7  0  0  8  4  0  9  1  1
REGI PQUD  4  0  9  0  0 10  6  0 11  1  1
REGI PQUD  6  0 11  0  0 12  5  0 13  1  1
MACR END 'TWO'
MACR ROLL 'TWO' 90
MACR REPL 'TWO' 3
MACR END 'ONE'
END JOB
```

FINEL Input File

Applying the Boundary Conditions

The boundary conditions can be applied using the FINEL file and will be transferred to the ABAQUS data file along with the geometry information. In this case it is required that the edges at the end of the tool clamped by a collet in the spindle are fixed. This is achieved in ABAQUS by defining node sets for each of the edges concerned and declaring them as fixed boundaries.

Multiple Point Constraints

Although the geometry can be defined using FINEL in terms of the regions shown in Figure 7, the elements used to construct the model are as shown in Figure 3. In ABAQUS these elements are classed as C3D20R (20 noded brick element with reduced integration scheme). Figure 10 shows a cross section illustrating the exact configuration of the nodes and element boundaries.

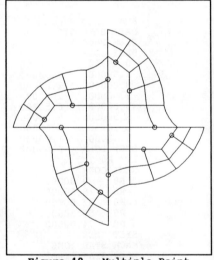

The generation of these elements in ABAQUS is straight forward – using the geometry generated by FINEL. However, Figure 10 shows that some elements have nodes that are not shared by the elements adjacent to them. In order to preserve the integrity of the model once loads are applied and deformations develop, we must define these nodes (those circled in Figure 10) to be constrained to the edges of the elements that they are adjacent to. This is achieved in ABAQUS by using the MCP - Multiple Point Constraint definition.

Figure 10 - Multiple Point Constraints

Applying the Loads

To simulate the cutting operation we can represent the loads on alternate edges of the cutting tool by defining amplitude variations as shown in Figure 11. As can be seen, the load varies

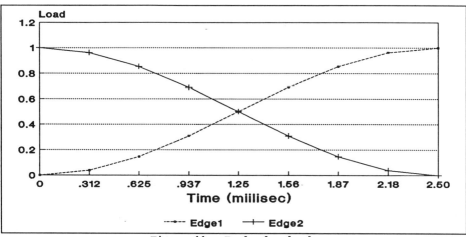

Figure 11 - Tool edge loads

on each cutting edge as cos 2θ, or cos 2(2πN/T)t, where the value
of T is determined by the spindle speed rate. In this case we
have applied a uniformly distributed edge load to the engaged
edge of the cutter - then through the appropriate loading
algorithm we have calculated the kinematic equivalent
concentrated point load and applied it to the tool.

The depth of the tool cut is defined by the number of edges that
have the load applied to them. The depth can therefore be
increased by applying the load to edges of adjacent elements. The
depth of cut can be decreased by altering the size of the
elements along the direction of the length of the tool.

Post-Processing

The solution of the model described above takes many hours on the
APOLLO DN10000 that we use at the Polytechnic. The output from
the analysis can be viewed using the FEMVIEW post-processor. This
package allows full colour graphical representations of stress
patterns to be displayed and Figures 12 and 13 are examples of
this form of output from FEMVIEW. Figure 12 shows contours of the
VonMises stress values calculated for the tool model and Figure
13 shows the Principle Shear calculated along one particular edge
of the helix. Plots like these from any point of view, can be

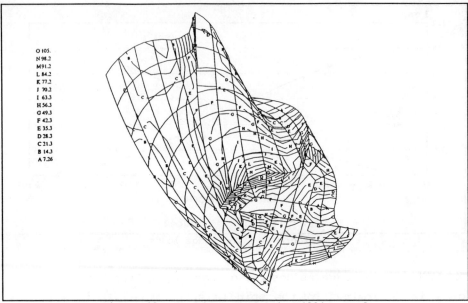

Figure 12 - VonMises Stress in Milling Tool

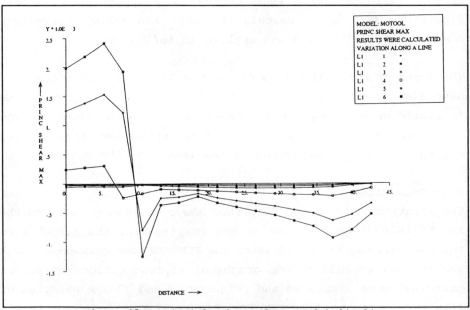

Figure 13 - Principle Shear along tool helix line

readily obtained from FEMVIEW the model.

Because of the amount of time that it takes to solve this model, the number of runs that each student can make is limited. However, by working in groups and by altering various parameters in the model results for various spindle speeds, depths of cut, helix angle, etc. can be obtained and compared.

CONCLUSIONS

We have seen how, at Leeds Polytechnic, we use integrated software packages to aid our 'black box' approach to the teaching of FEA. Students can use different packages to create geometry, perform the analysis and view the results. By using different examples - starting with very simple and building up to the more complex - we can cater for classes that have varying levels of experience and understanding. The example of the helical milling cutter is used to show how a reasonably complex model can be created and analysed in this way.

One of the FEA packages (ABAQUS) can also be used to solve more complex problems and many students have used this package for their final year projects (e.g. Jalaludin [2]). Other examples of industrial and academic projects exist in the Centre for Advanced Research in Engineering (CARE) at Leeds Polytechnic. There are written reports that outline these projects ([3]- [21]) and students can use these to help them with their work and/or to develop as further projects.

ACKNOWLEDGEMENTS

The authors would like to acknowledge the help and assistance given by the Centre for Advanced Research in Engineering (CARE) at Leeds Polytechnic.

REFERENCES

1.   Chatterton, S.A., FINEL Version 3.9 User Reference Manual, 1989.

2.   Jalaludin, A.H., Study of the Stress Field Pattern in Chain Links under Dynamic Loading, Final Year Project, Industrial & Manufacturing Division, Faculty of Information & Engineering Systems, Leeds Polytechnic, May 1991.

3.   Kormi, K., Islam, N.W., Simulation of Tyre Dynamics by Finite Element Techniques, <u>Applied Stress Mechanics</u>, ed. Hyde, T.M. and Ollerton E., Elsevier Applied Science, London, 1990.

4.   Kormi, K., Finite Element Analysis of Dynamic Buckling of a Cylindrical Shell with an Initially Induced Imperfection, Leeds Polytechnic, Centre for Advanced Research in Engineering, internal document, 1991.

5.   Kormi, K., Response of Structures to Impulse and Impact Loading, Leeds Polytechnic, Centre for Advanced Research in Engineering, internal document, 1990.

6.   Kormi, K. and Dudell, D. A., The Mushrooming of Flat, Ellipsoidal and Torispherical ended Projectiles Impinging on a Rigid Surface, Leeds Polytechnic, Centre for Advanced Research in Engineering, internal document, 1991.

7.   Kormi, K., Response of Wire Rope to Dynamic Loading Numerical Analysis, Leeds Polytechnic, Centre for Advanced Research in Engineering, internal document, 1990.

8.   Kormi, K. and Adams, D. R., Elastic Plastic Dynamic Response of Toroidal Chain Links to Triangular Pulse Type Loading, Leeds Polytechnic, Centre for Advanced Research in Engineering, internal document, 1991.

9.   Kormi, K. and Islam, M.N., Dynamic Response of an Anchored Buoyancy tank due to side pressure pulse, Leeds Polytechnic, Manufacturing Systems Engineering Department Internal Report, 1989.

10.   Islam, M.N., Kormi, K. and Al-Hassani, S.T.S, Dynamic Response of a thin walled cylinder to side pressure pulse, Leeds Polytechnic, Manufacturing Systems Engineering Department Internal Report, 1989.

11.   Islam, M.N., Kormi, K. and Al-Hassani, S.T.S, Analysis of Dynamic Response of offshore oil platform using Finite Element Method, Leeds Polytechnic, Manufacturing Systems Engineering Department Internal Report, 1989.

12.   Kormi, K., The Mushrooming of Flat-Ended Projectiles Impinging on a Rigid Surface, Leeds Polytechnic, Centre for Advanced Research in Engineering, internal document, 1991.

13.   Kormi, K., Dynamic expansion inhibited by the resistance of a fluid Medium of an enclosed cavity subjected to an internal pressure pulse, Leeds Polytechnic, Centre for Advanced Research in Engineering, internal document, 1991.

14.   Kormi K., Response of Structures to Impulse and Impact loading, in <u>FEM '90 Congress</u>, ed. Ikoss, 1990.

15.   Kormi, K. and Duddell, D.A., Response of structures to

impulse and impact loading II, Leeds Polytechnic, Centre for Advanced Research in Engineering, internal document, 1991.

16. Kormi, K. and Nikbin, K., Three dimensional modelling of an elliptical crack in a multi-layer composite, square box section cell, Leeds Polytechnic, Centre for Advanced Research in Engineering, internal document, 1991.

17. Kormi, K., Response of Newton's Cradle to Impact Loading, Leeds Polytechnic, Centre for Advanced Research in Engineering, internal document, 1990.

18. Kormi, K. and Islam, N.M., Simulation of Tyre Dynamics by Finite Element Techniques, in Applied Stress Mechanics, ed. Hyde, T.M. and Ollerton, E., Elsevier Applied Science, 1990.

19. Kormi, K., Shaghouei, E. and Duddell, D. A., Finite element examination of the dynamic response of clamped beam grillages impacted transversely by a rigid mass at their centres, Leeds Polytechnic, Centre for Advanced Research in Engineering, internal document, 1991.

20. Kormi, K. and Duddell, D.A., Response of structures to impulse and impact loading II, Leeds Polytechnic, Centre for Advanced Research in Engineering, internal document, 1991.

21. Kormi, K. and Islam, M.N., Dynamic Response of an Anchored Buoyancy Tank due to Side Pressure Pulse, Leeds Polytechnic, Manufacturing Systems Engineering Department. Internal Report, 1989.

DESIGN AND ANALYSIS OF A TRUCK SUSPENSION MOUNTING BRACKET

R. ALI and C. CHEUNG
University of Technology, Loughborough

ABSTRACT

Finite element analysis of a truck suspension mounting bracket is discussed. Several models of the bracket were developed and were analysed for a number of load cases. Some experimental validation was performed and is reported. It was found necessary to suggest design modifications to the bracket.

INTRODUCTION

The paper describes the finite element analysis and experimental validation of the design of a prototype commercial vehicle suspension mounting bracket. The project arose as a result of a period of industrial training spent by the co-author with a commercial vehicle manufacturer. During this period of training the student was involved in initial design of the bracket using classical techniques. Interest was expressed in the in depth analysis of this component using the finite element technique as a final year undergraduate degree project. This was acceptable to both the university and the industrial partner.

Stress analysis of the suspension bracket was carried out using the finite element method and four different road running conditions of the vehicle were considered for this analysis. These were acceleration, braking, cornering and shock loads. The results were partially validated by simulation and measurement of stresses in the laboratory. A comparison of theoretically predicted stresses and measured stresses is presented here.

As a result of this analysis, it was discovered that the design of the component was unsatisfactory in two regions and gave rise to high stress concentration. These regions of the bracket were redesigned in a manner so as to diffuse the stress. The redesigned bracket was analysed again with satisfactory results showing an improved factor of safety.

In addition to the discussion of the aforementioned analysis, an opportunity is taken to describe in some detail the discourse of structural mechanics in general and the finite element method in particular as part of our undergraduate course.

BACKGROUND EXPERIENCE AND LEARNING CURVE

The department of Transport Technology at the University of Technology, Loughborough, offers undergraduate courses in Aeronautical Engineering and Automotive Engineering. During the first two years of the course students are exposed to, among other subjects, fundamental principles of Solid Mechanics. During the third and final year of the course students select optional subjects. Two of the options available to the students are Stress and Structural Analysis and Computational Methods. Although it is possible to select either of these options individually, majority of students select both because of their interdependence.

The Stress and Structural Analysis option deals with topics such as theory of elasticity, theory of thin plates, torsion and warping, load diffusion, energy concepts and energy theorems. This latter topic is of considerable interest to students wishing to study the finite element method, covered under the heading of Computational Methods, as most of the development of the method is based on a good understanding of various energy theorems.

During the discussion of Computational Methods emphasis is placed on the use of matrix methods applied to structural analysis with the stiffness method as a preamble. The stiffness method introduces students to think in terms of matrix manipulation of quantities such as force and displacement vectors. This approach, subsequently, develops into the discussion of the finite element method with the students being exposed to various stages of its development such as idealisation of continua, various types of finite elements, development of stiffness matrices of various elements, coordinate transformation, transformation and assembly of matrices, application of boundary conditions and development of consistent load vectors. To grasp the mechanics of the system, students are encouraged to solve small problems manually following the same procedure as a machine would follow for the solution of larger problems.

In parallel with theoretical studies of the method students are required to become proficient in the use and application of a commercial finite element package.   They are taught to use

PAFEC as the analysis code and PIGS as the pre and post processor. The laboratory exercises are devised to demonstrate programming techniques such as use of symmetry and anti symmetry, application of mathematical boundary conditions and simulation of physical restraints. Normally exercises are chosen where classical results exist or can be easily evaluated to impart to the student added confidence in the analysis. A typical laboratory exercise would be the prediction of stress and deformation in a rotating disc.

To reinforce the learning curve and experience the students from this option are encouraged to undertake a project which is likely to extend their knowledge of this method. It is highly desirable, although not mandatory, that the results obtained from a finite element analysis are validated by appropriate experimental work in the laboratory. This provides the student with an opportunity to perform an in depth study of the technique itself, its application to industrial problems and its limitations. Above all the students gain experience in the interpretation of results from a finite element analysis. The project described in this paper is the result of such an exercise.

THEORETICAL MODELLING

The suspension mounting bracket is shown in Figure 1. It is a grey iron sand casting. The bracket is fixed to the chassis by five bolts which are tightened to a torque of 250 Nm. The main bolt carrying the spring element is tightened to a torque of 500 Nm.

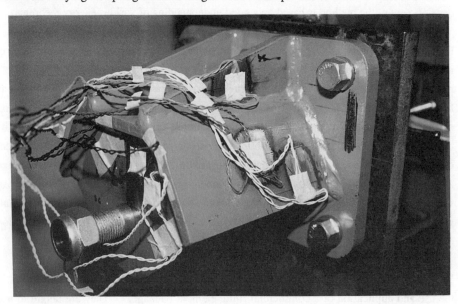

Figure 1.   Suspension mounting bracket.

It was decided to analyse the component for a number of standard design cases. The design cases considered and corresponding loads are given in table 1.

TABLE 1

Load Cases

Load Case	Node No	Global Direction	Load Magnitude kN
Acceleration	8, 36 46, 69	y -x	22.65 14.4
Cornering	8, 36 93, 5, 82 8, 4	y -z -z	11.2 3.2 1.6
Shock load	8, 36 5, 37	y x	22.25 42.95
Unit load	4, 38	-y	0.5

In view of these design cases, three types of finite elements can be used for modelling the bracket, thin shell, thick shell or solid brick elements. However the actual thickness of metal and geometry of the bracket precluded the selection of thin shell elements and the use of three dimensional elements was considered to be too expensive. Accordingly it was decided to use 8 noded quadrilateral and 6 noded triangular thick shell elements. A number of full and half models of the bracket were developed. One idealisation is shown in Figures 2 and 3. For symmetrical load cases (cornering) half models were utilised. For the purposes of this analysis some minor bolt holes were ignored and the main bolt holes were represented by square holes. Bosses round the holes were also ignored. Since the bolt location was ignored in the idealisation, constraints were applied to adjacent nodes. These are also shown in Figures 2 and 3. In view of the magnitude of the torque applied to the fixing bolts, it was considered that the fixing was sufficiently rigid to be totally constrained. The full model had 56 elements and 197 nodes. Analysis was carried out using PAFEC, level 6.2 and PIGS level 4.1 on a Prime 750 computer. Some of the predicted stresses are given in table 2.

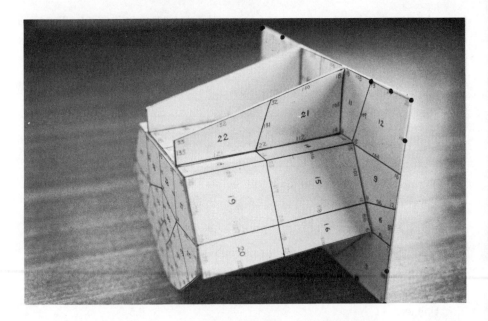

Figure 2.   Finite element model of the bracket - view 1.

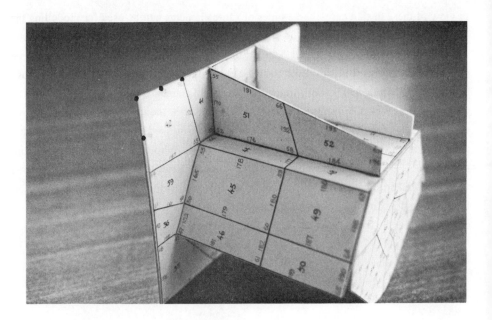

Figure 3.   Finite element model of the bracket - view 2.

EXPERIMENTAL ANALYSIS

A rig was designed to test the bracket in the laboratory and to verify the results obtained form the theoretical analysis. Due to the lack of available facilities it was not possible to simulate the service loads that would occur in practice. Instead the bracket was subjected to a simple tensile force while supported by a rigid column representing the chassis frame. This arrangement is shown in Figures 4 and 5. A number of 45° strain gauge rosettes were fixed to the bracket at strategic places and connected to a strain recorder. The location of the rosettes was arrived at by examining the stress predictions from the finite element analysis. These are depicted in figures 6 and 7. The bracket was loaded by a hydraulic pump and jack arrangement. The load was gradually increased in steps of 300 p.s.i. to a maximum value of 1500 p.s.i. and strain measurements were made using the strain recorder. The measured strain results are shown in Figure 8. The stresses measured from this load application exercise were compared with those predicted by the finite element models due to the application of a unit load. A comparison of measured and predicted stress results is given in table 2.

Figure 4.   Test rig.

Figure 5.   Test rig.

Figure 6.   Strain gauge rosettes 1 and 2.

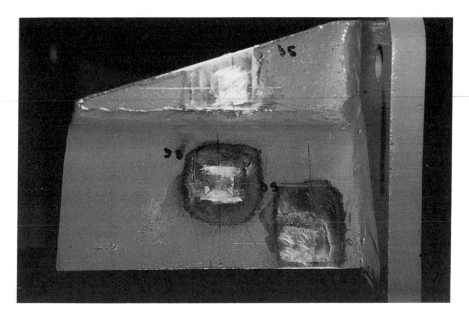

Figure 7.   Strain gauge rosettes 3, 4 and 5.

TABLE 2
Predicated and measured stresses

Rosette	Element No.	Node No.	σ_1		σ_2		τ	
			Predicted	Measured	Predicated	Measured	Predicted	Measured
1	28	28	-5.3	0.28	-8.7	-8.1	3.7	4.2
2	28	34	3.8	4.5	-5.1	-6.2	2.6	5.3
3	22	131	10.8	12.4	0.17	0.15	5.3	6.2
4	19	117	3.5	4.2	-7.2	-5.9	5.4	5.1
5	16	16	0.8	0.4	-6.4	-9.8	3.6	4.7

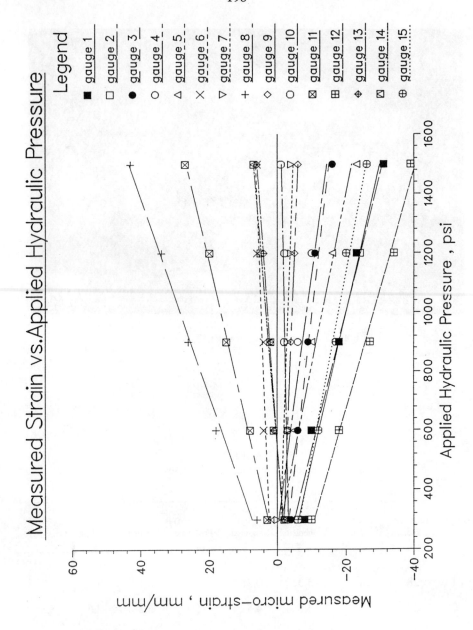

Figure 8a

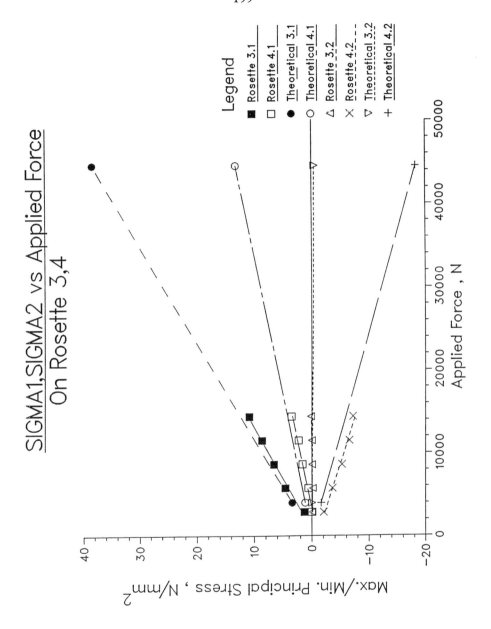

SIGMA1,SIGMA2 vs Applied Force
On Rosette 3,4

Figure 8b

DISCUSSION OF RESULTS

The correlation between predicted and measured stress results for rosette gauge positions 3 and 4 was found to be satisfactory. Agreement between the two set of results is within 20%. However the stress results corrsponding to the position of gauge 5 were not very encouraging. The error between the theoretical and measured stress values at gauge number 5 was 53%. This discrepancy can be explained by the fact that at element number 16, the stress graient is very high and changing rapidly. The rosette strain gauge being of a finite length can only record average strain while the predicted stress is the average over an element which is comparatively large. Also it was noticed that the gauge was not precisely located at the centre of the element. Also relatively large discrepancy in stress figures at the location of gauges 1 and 2 is conisdered to be caused by the geometry of the component. These results can be improved if solid finite elements were to be used.

From this initial work it was concluded that the type of elements used and the finite element mesh were reasonably satisfactory to proceed with the analysis of the suspension mounting bracket for the design load cases given in table 1. Substructuring could be used to study more closely areas of interest in subsequent runs. All design load cases were then analysed using the above finite element moels. Some of the predicted stress values for these load cases are given in table 3.

TABLE 3
Critical stress values

Load Case	Element No.	Node No.	Stress N/mm^2
Acceleration	46	47	120.0
Acceleration	51	55	-23.5
Cornering	16	6	56.7
Cornering	21	18	-14.6
Shock load	16	6	326
Shock load	59	48	238
Shock load	21	18	273

It was evident from the analysis that the shock load was the most critical load case and that under this condition, the component is likely to suffer failure. The problem areas are shown in Figure 9 and 10. Fatigue damage is also likely to be caused in this area. To strengthen the section in this area and to diffuse the stress, lips were introduced in the model as shown in Figure 11 and 12. As a result of introduction of lips on the edges of the suspension

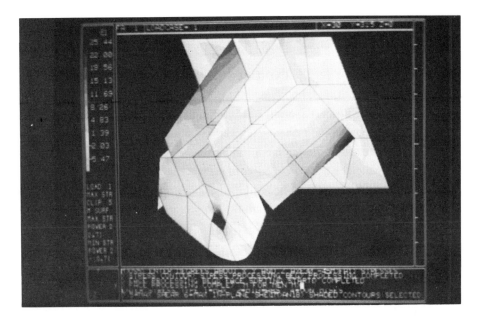

Figure 9.   Maximum principal stress - shock load.

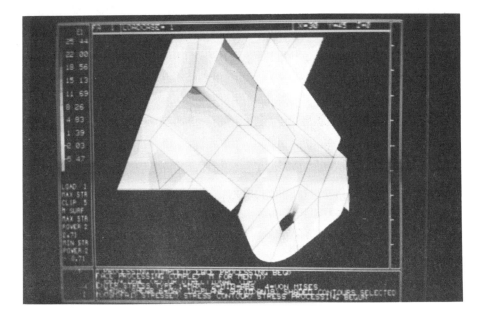

Figure 10. Maximum principal stress - shock load.

mounting bracket, the principal stresses were reduced by 27%, for the shock load case, 48% for the cornering load case and 40% for the acceleration load case.

CONCLUSIONS

As a result of the finite element analysis discussed in this paper it was concluded that the current design of the suspension mounting bracket was unsatisfactory and is likely to fail if subjected to shock loads. A design modification was suggested. This consisted of the provision of a 5mm high lip at the front edges of the bracket. It has been shown that if this suggested design modification is implemented, the component is likely to provide trouble free service.

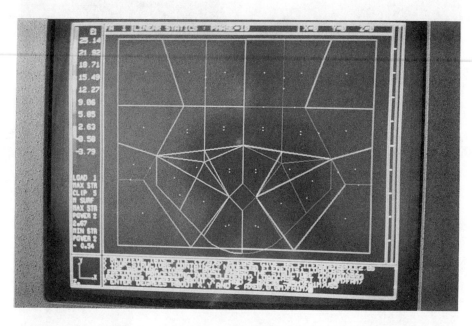

Figure 11. Modified model mesh - view 1.

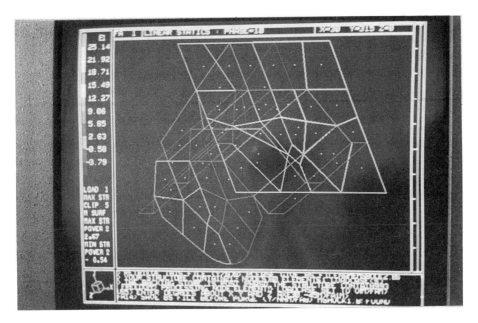

Figure 12. Modified model mesh - view 2.

EXPERIENCES IN USING THE SWEDISH TEACHING PACKAGE CALFEM

J R RIDDINGTON and J D RICHARDSON

School of Engineering, University of Sussex,

Falmer, Brighton, BN1 9QT

ABSTRACT

The Swedish teaching package CALFEM* is not a conventional finite element package. It is an interactive computer program based on a command language. The user controls the computation process by issuing commands which effectively call up subroutines. To use the program, the user has to understand the stages involved in the solution process. CALFEM is currently used in two undergraduate courses at the University of Sussex. In the first, it is used to aid the teaching of the stiffness method. It has proved to be effective in teaching the basic stages involved in undertaking an analysis. In the second course, it is used to aid the teaching of the finite element method. Although it has been shown to be effective in demonstrating the stages involved in an analysis, it has been found to be too cumbersome to use for anything but the simplest of problems.

INTRODUCTION

In the teaching of finite elements, it is generally felt that giving students hands-on experience of real commercial software packages is

---

*CALFEM is an acronym for 'computer-aided learning of the finite element method.' The package was produced in 1986 by: The Division of Structural Mechanics, Lund Institute of Technology, Lund, Sweden.

important. However, operating such packages without at least some theoretical understanding of finite elements is not to be recommended and most finite element courses spend the majority of time providing a firm analytical base. Thus, there are generally two basic requirements for any finite element course: to give students

> (i) a suitable theoretical grounding in
> the finite element method.

> (ii) practical experience of using one or
> more commercial packages.

Unfortunately, though, the running of commercial packages does not reinforce directly the theoretical knowledge gained earlier in a course, so there is little overlap between these two requirements. There is also in many courses the added problem of assessment: does one assess the theoretical understanding of the subject or simply a student's ability to assemble a data file, in which case should a design element be built into any assessment exercise?

Until recently at Sussex, students were required to write their own Fortran programs each of which obviously had to be severely limited in scope - applicable to a particular class of element and for a restricted range of problem e.g. three-node axi-symmetric elements for stress analysis problems with no more than 50 elements, 50 nodes and 3 load cases. This was an attempt to bridge the gap between the two requirements referred to above as well as providing a good means of assessment. Latterly, though, the teaching package CALFEM has been used because. in order to operate this package. the user needs to understand what is happening and what is required at each and every stage of the computation.

CALFEM is an interactive computer program which can be used to analyse a range of different types of structural mechanics and field problems. Since it is intended for the express purpose of teaching the principles of the finite element method, it is necessarily cumbersome to use and clearly cannot realistically be employed in applications where large numbers of elements are required. It is based on a command language and the computation is dictated directly by the user. Each command effectively calls up a subroutine and each command consists of a command name followed by a number of arguments - generally names of matrices. For example, command ELIN EN EK K reads element stiffness matrix EK into structure stiffness matrix K at positions defined by topology matrix EN.

As mentioned above, inputting data is cumbersome but necessarily so if the user is to understand and follow the progress of the computation. Each element is considered in turn with its nodal coordinates input (for all but the simplest bar and beam elements) prior to CALFEM calculating the corresponding element stiffness matrix. Thus, there is considerable duplication of effort with most nodal coordinates being specified more than once in the 'command bank' (combined instruction and data file). In addition, for each element there is a command required to calculate the element stiffness matrix and another command to insert the matrix in the correct locations of the structure stiffness matrix. A further consideration is that with CALFEM, numbers representing all the nodal freedoms of each element are input rather than simply inputting node numbers and, although this makes combining different element types somewhat easier, the amount of data read into the topology matrix is much increased.

In spite of the unwieldy nature of the data input, the labour involved is not generally excessive for systems composed of line elements (bars and beams) but clearly only a small number of two-dimensional elements can realistically be considered. This limitation with regard to two-dimensional elements could be overcome to a certain extent by considering problems in which there was a small number of element families with all the elements in any given family having the same dimensions and orientation. and hence the same stiffness matrix.

A further consideration with regard to the number of elements is that, if the number is large, the commands can become repetitive and students could then add to the 'command bank' in a routine unthinking way which negates the whole teaching purpose of the CALFEM package. It may, therefore, be no bad thing for students to be given somewhat unreal problems employing only a small number of elements if they are then able to focus more on the mechanics of the computation process.

The range of elements available within CALFEM is limited to one and two-dimensional elements. The highest order elements are a four-node quadrilateral and a four-node rectangular plate-bending element. Non-linear problems generally cannot be dealt with although geometrical non-linearity can be allowed for in structures consisting entirely of bar and/or beam elements. Likewise, dynamics problems can be considered but, with the exception of eigenvalue problems, only for structures consisting of beam elements. Finally, transient analysis of such problems as thermal conduction and groundwater flow are available using three-node triangular elements.

One drawback not yet mentioned is the lack of graphical facilities with CALFEM. Not being able to draw the mesh or to display the computed results is clearly a major disadvantage.

PRESENT USE OF CALFEM IN NUMERICAL STRUCTURAL ANALYSIS COURSE

The following three CALFEM exercises were set as part of a second year numerical structural analysis course. This course covers both the flexibility and stiffness methods. However, the major part of the course is devoted to the stiffness method and all the CALFEM exercises use this approach.

Exercise 1

In the first exercise, nodal displacements have to be calculated for the simple one dimensional structure shown in Fig 1 for a zero and a prescribed deflection at node 1. This introductory exercise is intended to ensure students understand the basic stages involved in a simple stiffness analysis as well as some different matrix solution techniques. The problem is a lecture example which occurs early in the course and the students are required to enter hand calculated stiffness and equivalent nodal force matrices rather than calculate these within CALFEM. The following three different equation solution techniques have to be employed:

  (i)   use of the standard CALFEM equation solution
        routine SOLVE

  (ii) matrix partitioning prior to inversion of the
        reduced stiffness matrix using CALFEM's routine INV

  (iii) modification of the stiffness matrix and the
        overall force matrix to account for prescribed
        displacements prior to inversion of complete
        modified stiffness matrix.

A CALFEM command bank listing of a solution for this problem when there is a prescribed support displacement is shown in Appendix 1.

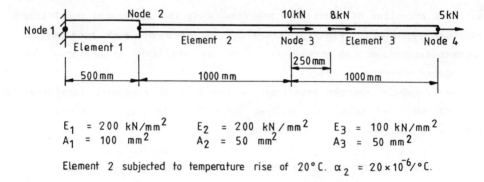

$E_1 = 200$ kN/mm^2 $E_2 = 200$ kN/mm^2 $E_3 = 100$ kN/mm^2

$A_1 = 100$ mm^2 $A_2 = 50$ mm^2 $A_3 = 50$ mm^2

Element 2 subjected to temperature rise of 20°C. $\alpha_2 = 20 \times 10^{-6}/°C$.

Figure 1 First CALFEM Exercise

Exercise 2

In the second exercise, students are required to undertake a
stiffness analysis of the simple two dimensional structure shown in Fig.
2. CALFEM has a number of different two-node line elements available and
this analysis requires the use of the two dimensional pin ended bar
element and the two dimensional frame type element. This exercise
extends the application of the stiffness method from the one dimensional
case considered in the first exercise to a more practical type of
application and it uses more of CALFEM's facilities. All of the matrices
required in the analysis are formed from basic dimensions and load data
rather than being calculated by hand and then read in as data, as was the
case with the first exercise. This example also illustrates the factors
that have to be taken into consideration when undertaking an analysis
which involves the combination of elements of different types. A CALFEM
command bank listing of a solution for this problem is shown in Appendix
2. To illustrate what appears on the computer screen whilst running
CALFEM, which can also be saved if required in a listing bank, the
contents of the listing bank produced when running the command bank is
shown in Appendix 3

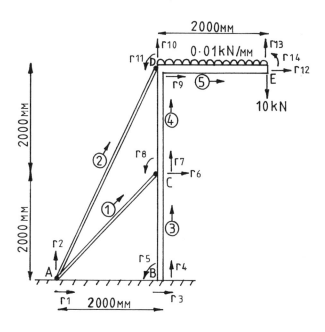

Figure 2. Second CALFEM Exercise

Exercise 3

The final exercise given to the students in the Numerical Structural Analysis course is a dynamics problem. The first four flexural natural frequencies and modes of vibration have to be calculated for the circular shaft and pulley system shown in Fig 3 using a lumped mass approach. Dynamics was introduced into the course in order to illustrate the fact that the stiffness method has wider applications than simply the static analysis of frame and beam structures. CALFEM has built in routines which allow the solution of this type of problem to be easily accomplished. The DIAG command allows a diagonal lumped mass matrix to

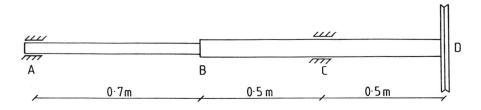

Figure 3. Third CALFEM Exercise

be formed easily and the EIGEN command produces a solution for eigenvalues and eigenvectors once the structure stiffness and lumped mass matrices have been formed. A CALFEM command bank listing of a solution for this problem using 17 elements is shown in Appendix 4.

PRESENT USE OF CALFEM IN NUMERICAL STRESS ANALYSIS COURSE

Currently CALFEM is used for an introductory and an assessment exercise in a final year numerical stress analysis course. This year's assessment exercise was to analyse the symmetric plane stress structure shown in Fig 4 using

(i) three node triangular elements,

(ii) four node quadrilateral elements and

(iii) eight node isoparametric element

in order to compare the accuracy of the results obtained. A separate program had to be used for the eight node isoparametric analysis because CALFEM's element library does not contain this element type. CALFEM was found to be not well suited to an analysis of this type where virtually every element is a different size. The command bank file becomes

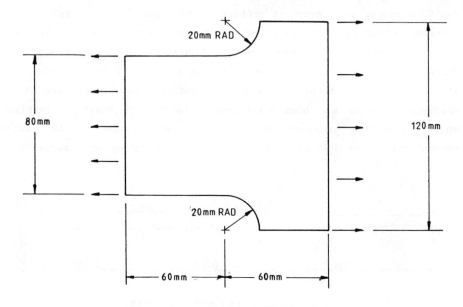

Figure 4. Fourth CALFEM Exercise

excessively long because seven lines of input are required to form each quadrilateral element stiffness matrix, three lines of input are required to read each element stiffness matrix into the structure stiffness matrix and each element has to be called up separately to obtain stress results. For bodies which can be sensibly divided into elements with a limited number of different shapes, such as that shown in Fig 5, the number of lines of input required by CALFEM is greatly reduced and any future finite element exercise involving CALFEM will be of this type. Because of the length of the command bank required to undertake this exercise, only part of it is shown in Appendix 5.

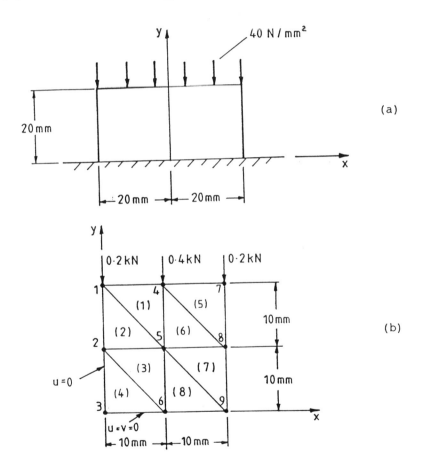

Figure 5. Suitable element grid to use with CALFEM

DISCUSSION

Overall it has been found that CALFEM has been more useful for teaching the stiffness method than the finite element method. It has been successful in ensuring that students fully understand the basic stages involved in undertaking a stiffness analysis. Being able to include a dynamics exercise has also demonstrated to students that the methods have wider applications outside the immediate areas of force and stress analysis. This eigenvalue problem with the accompanying theory covered in the lecture course, has also been found to provide a good introduction to a structural vibrations course taught in the following term. CALFEM does, however, have very limited capabilities even with regard to the stiffness method. These include its restriction to one loadcase and its inability to cope with initial strains unless the initial strain matrices are calculated by hand. It has no graphical display facilities and is not practical to use for analysing large structures which require the use of many elements. CALFEM also does not use a frontal type solution.

For teaching the finite element method, it is felt to be less useful and the limitations of the package more significant. It is considered that it is really only of value for demonstrating the basic steps involved in an analysis and the similarity between the stiffness and finite element methods. It is not a suitable package to use for investigating factors such as the accuracy of different element types and the importance of the element grid pattern because of the number of lines of input required. Not having the commonly used eight node isoparametric element available is also a disadvantage. In future, it is envisaged that CALFEM will be employed in any finite element course merely as a way of introducing the subject and to highlight the similarities between the finite element and stiffness methods. However, a different package with simple data input and good graphical output will be used in any subsequent assessment exercises.

CONCLUSIONS

In teaching the stiffness method within numerical structural analysis courses, CALFEM can play a valuable role in providing a bridge between the theoretical foundations of the subject and the use of commercial packages. In a finite element course, however, CALFEM is much more limited in the range of problems that can sensibly be attempted. If realistic problems are considered with an adequate number of elements, CALFEM will not only be cumbersome to use but will tend to be used in a routine way thus negating its teaching purpose. Nevertheless, CALFEM has a role to play, even if problems are inevitably somewhat artificial, particularly in teaching the basic principles of the finite element method and emphasising similarities with the stiffness method.

APPENDIX 1

```
C ASSIGNMENT EXERCISE 1 : FIXED DISPLACEMENT APPLIED
C 3 DIFFERENT SOLUTION METHODS USED
C
C FORM A ZEROED STRUCTURE STIFFNESS MATRIX K
LMAT K 4 4
ZERO
C FORM ELEMENT STIFFNESS MATRICES
LMAT K1 2 2
40 -40
-40 40
LMAT K2 2 2
10 -10
-10 10
LMAT K3 2 2
5 -5
-5 5
C FORM ELEMENT EQUIVALENT NODAL FORCE MATRICES
C FOR NON NODAL LOADING
LMAT FD1 2 1
0
0
LMAT FD2 2 1
0
0
LMAT FD3 2 1
6
2
C FOR THERMAL STRAINS
LMAT FE1 2 1
0
0
LMAT FE2 2 1
-4
4
LMAT FE3 2 1
0
0
C FORM MATRIX OF FORCES APPLIED AT NODES R
LMAT R 4 1
0
0
10
5
C FORM ZEROED STRUCTURE EQUIVALENT FORCE MATRICES
LMAT FD 4 1
ZERO
LMAT FE 4 1
ZERO
C LOAD ELEMENT STIFFNESS AND EQUIVALENT NODAL FORCE MATRICES
C INTO STRUCTURE STIFFNESS AND FORCE MATRICES
ELIN CON1 K1 K FD1 FD FE1 FE
1
1 1 2
ELIN CON2 K2 K FD2 FD FE2 FE
1
1 2 3
ELIN CON3 K3 K FD3 FD FE3 FE
1
1 3 4
C FORM OVERALL STRUCTURE FORCE MATRIX F
ADD FD FE FDE
ADD FDE R F
```

```
C SOLUTION METHOD 1 : USING SOLVE COMMAND
C
SOLVE K U F B Q
1
1 2
C PRINT NODAL DISPLACEMENT SOLUTION
PRINT U
C PRINT NODAL RESTRAINT REACTIONS
PRINT Q
C
C SOLUTION METHOD 2 : PARTITIONING TO REDUCE THE NUMBER OF EQUATIONS
C
C FORM REDUCED SIZE STRUCTURE STIFFNESS MATRIX
REDUCE K KK C
1
1
C FORM ZEROED REDUCED SIZE NODAL FORCE MATRIX FSEG
LMAT FSEG 3 1
ZERO
C COPY SEGMENT OF F INTO FSEG
COPY F FSEG 1 1 2 1 4 1
C DEFINING FIXED DISPLACEMENT
LMAT DISP
2
C FORM ZEROED COLUMN MATRIX F1
LMAT F1 3 1
ZERO
C COPY COLUMN OF K RELATING TO FIXED DISPLACEMENT INTO F1
COPY K F1 1 1 2 1 4 1
C MULTIPLY F1 BY FIXED DISPLACEMENT
MULT F1 DISP F2
C SUBTRACT F2 FROM FSEG TO FORM FINAL MODIFIED FORCE MATRIX FF
SUB FSEG F2 FF
C INVERT REDUCED SIZE STRUCTURE STIFFNESS MATRIX
INV KK KKINV
C MULTIPLY INVERTED MATRIX BY FORCE MATRIX TO GET DISPLACEMENTS
MULT KKINV FF UU
C PRINT OUT DISLACEMENT SOLUTION
C MATRIX ROW 1 CORRESPONDS TO NODE2, ROW 2 TO NODE 3 ETC.
PRINT UU
C
C SOLUTION METHOD 3 : MODIFYING STIFFNESS AND FORCE MATRICES
C
C DEFINE VERY-LARGE NUMBER
LMAT NN
1E10
C MODIFYING STIFFNESS MATRIX
LMAT KTERM
0
COPY K KTERM 1 1 1 1 1 1
ADD NN KTERM KMOD
COPY KMOD K 1 1
C DEFINING FIXED DISPLACEMENT
LMAT DISP
2
C MULTIPLYING FIXED DISPLACEMENT BY NEW DIAGONAL TERM IN K
MULT KMOD DISP FMOD
C REPLACING ORIGINAL FORCE MATRIX TERM WITH FMOD
COPY FMOD F 1 1
C INVERTING MODIFIED STIFFNESS MATRIX
INV K KINV
C MULTIPLYING INVERTED K BY MODIFIED FORCE MATRIX
MULT KINV F UUU
C PRINTING OUT SOLUTION FOR DISPLACEMENTS
PRINT UUU
C STORING LISTING BANK IN A FILE
LBSEND EX1.LB

ASSIGNMENT EXERCISE 1
```

APPENDIX 2

```
C ASSIGNMENT EXERCISE 2
C
C FORM ZEROED STRUCTURE STIFFNESS MATRIX K
LMAT K 14 14
ZERO
C FORM MATRIX OF FORCES APPLIED AT NODES
LMAT R 14 1
GO TO 13
-10
0
C CREATE ELEMENT STIFFNESS MATRICES AND ADD INTO MATRIX K
C FOR ELEMENT 5 ALSO FORM MATRIX OF NODAL FORCES EQUIVALENT
C TO DISTRIBUTED LOADING AND ADD INTO MATRIX R
BAR2E EP1 EK1
2828 45 200 200
ELIN EN1 EK1 K
1
1 1 2 6 7
BAR2E EP2 EK2
4472 63.43 200 200
ELIN EN2 EK2 K
1
2 1 2 9 10
BEAM2E EP3 EK3
2000 90 2000 3E6 200
ELIN EN3 EK3 K
2
3 3 4 5 6 7 8
4 6 7 8 9 10 11
BEAM2E EP5 EK5 ER5
2000 0 8000 10E6 200
-0.01
ELIN EN5 EK5 K ER5 R
1
5 9 10 11 12 13 14
C SOLVE SYSTEM OF EQUATIONS, ENTERING PRESCRIBED DISPLACEMENTS
SOLVE K U R B
5
1 0
2 0
3 0
4 0
5 0
C PRINT DISPLACEMENT SOLUTION
PRINT U
C CALCULATE STRESS IN BAR ELEMENTS 1 & 2
BAR2S EP1 EN1 U
BAR2S EP2 EN2 U
C CALCULATE FORCES AND MOMENTS IN BEAM ELEMENTS
BEAM2S EP3 EN3 U
BEAM2S EP5 EN5 U
C STORING LISTING BANK IN A FILE
LBSEND EX2.LB
ASSIGNMENT EXERCISE 2
```

APPENDIX 3

££
£   CALFEM - Version 85:02-110   £
£   (C) Div of Struct Mechanics, Lund University, Sweden   £
£   OUTPUT FROM CALFEM   £
£   LIST IDENTITY: ASSIGNMENT 2   £
££

```
-----------------------------------
**CBON EX2.CB
 54 COMMAND LINES ARE STORED
-----------------------------------
**RUN
-----------------------------------
**C ASSIGNMENT EXERCISE 2
-----------------------------------
**C
-----------------------------------
**C FORM ZEROED STRUCTURE STIFFNESS MATRIX K
-----------------------------------
**LMAT K 14 14
 14 ROWS    14 COLUMNS IN MATRIX K
INPUT OF ROW NO   1 !
**ZERO
ZERO MATRIX
-----------------------------------
**C FORM MATRIX OF FORCES APPLIED AT NODES
-----------------------------------
**LMAT R 14 1
 14 ROWS     1 COLUMNS IN MATRIX R
INPUT OF ROW NO   1 !
**GO TO 13
INPUT OF ROW NO  13 !
**-10
INPUT OF ROW NO  14 !
**0
-----------------------------------
**C CREATE ELEMENT STIFFNESS MATRICES AND ADD INTO MATRIX K
-----------------------------------
**C FOR ELEMENT 5 ALSO FORM MATRIX OF NODAL FORCES EQUIVALENT
-----------------------------------
**C TO DISTRIBUTED LOADING AND ADD INTO MATRIX R
-----------------------------------
**BAR2E EP1 EK1
DEFINE LENGTH, SLOPE, AREA AND ELASTIC MODULUS!
**2828 45 200 200
-----------------------------------
**ELIN EN1 EK1 K
SPECIFY NUMBER OF IDENTICAL ELEMENTS!
**1
GIVE ELEMENT NUMBER AND    4 GLOBAL NODE VARIABLE NUMBERS!
**1 1 2 6 7
```

```
------------------------------------
**BAR2E EP2 EK2
DEFINE LENGTH, SLOPE, AREA AND ELASTIC MODULUS!
**4472 63.43 200 200
------------------------------------
**ELIN EN2 EK2 K
SPECIFY NUMBER OF IDENTICAL ELEMENTS!
**1
GIVE ELEMENT NUMBER AND   4 GLOBAL NODE VARIABLE NUMBERS!
**2 1 2 9 10
------------------------------------
**BEAM2E EP3 EK3
DEFINE LENGTH, SLOPE, AREA, MOMENT OF INERTIA AND ELASTIC MODULUS!
**2000 90 2000 3E6 200
------------------------------------
**ELIN EN3 EK3 K
SPECIFY NUMBER OF IDENTICAL ELEMENTS!
**2
GIVE ELEMENT NUMBER AND   6 GLOBAL NODE VARIABLE NUMBERS!
(NO    1 OF    2 )
**3 3 4 5 6 7 8
(NO    2 OF    2 )
**4 6 7 8 9 10 11
------------------------------------
**BEAM2E EP5 EK5 ER5
DEFINE LENGTH, SLOPE, AREA, MOMENT OF INERTIA AND ELASTIC MODULUS!
**2000 0 8000 10E6 200
DEFINE LOAD INTENSITY!
**-0.01
------------------------------------
**ELIN EN5 EK5 K ER5 R
SPECIFY NUMBER OF IDENTICAL ELEMENTS!
**1
GIVE ELEMENT NUMBER AND   6 GLOBAL NODE VARIABLE NUMBERS!
**5 9 10 11 12 13 14
------------------------------------
**C SOLVE SYSTEM OF EQUATIONS, ENTERING PRESCRIBED DISPLACEMENTS
------------------------------------
**SOLVE K U R B
GIVE NUMBER OF PRESCRIBED VALUES!
**5
GIVE ROW NUMBER AND CORRESPONDING VALUE!
(NO    1 OF    5 )
**1 0
(NO    2 OF    5 )
**2 0
(NO    3 OF    5 )
**3 0
(NO    4 OF    5 )
**4 0
(NO    5 OF    5 )
**5 0
------------------------------------
**C PRINT DISPLACEMENT SOLUTION
------------------------------------
**PRINT U
```

```
MATRIX U              14 ROWS    1 COLUMNS
=============================================

              1
     1  0.000000E+000
     2  0.000000E+000
     3  0.000000E+000
     4  0.000000E+000
     5  0.000000E+000

     6 -3.46176
     7-0.246839
     8  0.726741E-002
     9  13.9916
    10-0.624816

    11-0.500571E-001
    12  13.9916
    13 -124.072
    14-0.667238E-001
------------------------------------
**C CALCULATE STRESS IN BAR ELEMENTS 1 & 2
------------------------------------
**BAR2S EP1 EN1 U
ELEMENT NO   1      SIG =-0.185458        ( N = -37.0916       )
------------------------------------
**BAR2S EP2 EN2 U
ELEMENT NO   2      SIG = 0.254897        ( N =  50.9795       )
------------------------------------
**C CALCULATE FORCES AND MOMENTS IN BEAM ELEMENTS
------------------------------------
**BEAM2S EP3 EN3 U

ELEMENT NO   3
N1  = -49.3677      V1  = -3.42508        M1  = -1244.86
N2  = -49.3677      V2  = -3.42508        M2  =  5605.30

ELEMENT NO   4
N1  = -75.5955      V1  =  22.8027        M1  =  5605.30
N2  = -75.5955      V2  =  22.8027        M2  = -40000.0
------------------------------------
**BEAM2S EP5 EN5 U

ELEMENT NO   5
N1  = 0.000000E+000  V1  = -30.0000       M1  = -40000.0
N2  = 0.000000E+000  V2  = -10.0000       M2  = 0.244427E-011
------------------------------------
**C STORING LISTING BANK IN A FILE
------------------------------------
**LBSEND EX2.LB
GIVE LIST IDENTITY!
**ASSIGNMENT 2

££££££££££££££££££££££££££££££££££££££££££££££££££££££
£  END OF OUTPUT                                    £
£  LIST IDENTITY: ASSIGNMENT 2                      £
££££££££££££££££££££££££££££££££££££££££££££££££££££££
```

APPENDIX 4

```
C ASSIGNMENT EXERCISE 3
C
C FORM A ZEROED STRUCTURE STIFFNESS MATRIX K
LMAT K 36 36
ZERO
C FORM ELEMENT STIFFNESS MATRICES AND REMOVE AXIAL STIFFNESS TERMS
BEAM2E EP1 EK1
0.1 0 3.1416E-4 7.8539E-9 200E9
REDUCE EK1 EK11 C1
2
1 4
BEAM2E EP2 EK2
0.1 0 7.0685E-4 39.760E-9 200E9
REDUCE EK2 EK22 C2
2
1 4
C LOAD ELEMENT STIFFNESS MATRICES INTO STRUCTURE STIFFNESS MATRIX
ELIN EN1 EK11 K
7
1 1 2 3 4
2 3 4 5 6
3 5 6 7 8
4 7 8 9 10
5 9 10 11 12
6 11 12 13 14
7 13 14 15 16
ELIN EN2 EK22 K
10
8 15 16 17 18
9 17 18 19 20
10 19 20 21 22
11 21 22 23 24
12 23 24 25 26
13 25 26 27 28
14 27 28 29 30
15 29 30 31 32
16 31 32 33 34
17 33 34 35 36
C FORM DIAGONAL LUMPED MASS MATRIX
LMAT M 36 36
DIAG
0.1233 1E-11 0.2466 1E-11 0.2466 1E-11 0.2466 1E-11 0.2466 1E-11
0.2466 1E-11 0.2466 1E-11 0.4007 1E-11 0.5549 1E-11 0.5549 1E-11
0.5549 1E-11 0.5549 1E-11 0.5549 1E-11 0.5549 1E-11 0.5549 1E-11
0.5549 1E-11 0.5549 1E-11 5.2775 0.03
C SOLVE FOR EIGENVALUES AND EIGENVECTORS, ENTERING RESTRAINTS
EIGEN K M L X B
2
1 25
C PRINT SOLUTION FOR EIGENVALUES
PRINT L
C PRINT SOLUTION FOR FIRST FOUR EIGENVECTORS
PRINT X 1 1 36 4
C STORING LISTING BANK IN A FILE
LBSEND EX3.LB
ASSIGNMENT EXERCISE 3
```

APPENDIX 5

```
C ASSIGNMENT EXERCISE 4
C
C FORM A ZEROED STRUCTURE STIFFNESS MATRIX K
LMAT K 42 42
ZERO
C FORM MATRIX OF FORCES APPLIED AT NODES
LMAT R 42 1
-600
0
-1200
0
-600
0
GO TO 25
700
GO TO 31
1200
GO TO 37
500
ZERO
C FORM ELEMENT STIFFNESS MATRICES AND LOAD INTO K
PLANQE EP1 EK1
1
0 0
30 0
30 20
0 20
2
200000 0.3
ELIN EN1 EK1 K
1
1 1 2 7 8 9 10 3 4
      *
      *
      *
PLANQE EP12 EK12
1
78.478 52.346
95 45
95 60
80 60
2
200000 0.3
ELIN EN12 EK12 K
1
12 35 36 33 34 39 40 41 42
C SOLVE SYSTEM OF EQUATIONS, ENTERING PRESCRIBED DISPLACEMENTS
SOLVE K U R B
5
2 0
8 0
14 0
20 0
26 0
C PRINT DISPLACEMENT SOLUTION
PRINT U
C CALCULATE ELEMENT STRESSES
PLANQS EP1 EN1 U
        *
        *
        *
PLANQS EP12 EN12 U
C STORING LISTING BANK IN A FILE
LBSEND EX4.LB
ASSIGNMENT EXERCISE 4
```

TEACHING FINITE ELEMENTS TO DISADVANTAGED STUDENTS

Peter Utting
Professor of Structural Engineering
University of Natal

ABSTRACT

Students from a disadvantaged background - whether from a deprived education system or where engineering is perceived to be a male domain - often suffer from a deficiency in spatial perception which demands a physical rather than a mathematical approach to teaching structural behaviour.

This paper highlights the value of a quality finite element package as a vehicle for remedying spatial inadequacy by developing an instinctive feeling for structural behaviour. It argues that little, if anything, is to be gained by continuing to teach analytical techniques that were in fashion before computers came to be widely used by engineers. Moreover, it contends that structural engineering students should not have to undergo extensive programming courses so that they can develop a program to exhibit structural behaviour. Students need to be "functionally" computer literate so that they can use computers effectively.

Experience shows that spatial perception and computer usage are better understood by students, both disadvantaged and advantaged alike, when they are given hands on practice with a quality finite element package, preferably limited to about 50 elements. Problem specifications are provided for each assignment, together with an outline of the computer operations (commands and mouse movements) to achieve a solution. The students work individually on each assignment, for up to two weeks, and a report is submitted on completion. These reports are marked within a week to ensure that the students receive the feedback to enable them to improve as the course proceeds. Students make progress quickly, and even those who have had limited computer experience rapidly display computer proficiency. Continuous assessment of assignments and a final examination make up the total mark for the course. Significantly, there is no correlation between achievement levels and the gender or educational background of students.

INTRODUCTION

Nearly two years ago I arrived in South Africa at the dawn of the De Klerk era to teach structural engineering at the University of Natal. I had not been an academic for many years, and was thrown in at the deep end to teach not only a finite element course at both undergraduate and post-graduate levels, but also courses in basic structures and frame analysis. This paper is the result of that experience. It should be stated at the outset that the purpose is not to present research findings. Rather it is to promote discussion on the advantages of teaching finite elements to overcome the effects of a deprived and discriminatory educational background.

The first part of the paper gives an overview of the South African education system, and the disadvantages suffered by engineering students as a result of it. The second part addresses remedies. It looks at the need to reorganise courses to overcome the deficiencies resulting from a disadvantaged educational background. It considers the use of a spatial perception approach, and the necessity to develop what I term "functional computer literacy." The importance of planned assignments is discussed particularly with regard to working with physical experiments, known solutions and benchmarks. The assessment of students is evaluated in terms of continuous assessment and examinations.

THE DISADVANTAGED AND THE SOUTH AFRICAN EDUCATION SYSTEM

In 1989, South African educationist Ken Hartshorne made the prophetic statement that: "It will be more difficult for the government to change the principle of segregated education systems than to free Nelson Mandela, negotiate with the African National Congress or repeal the Group Areas Act, because [the principle] lies at the heart of the ideology of separate development."[1]

Even though pillars of apartheid such as the Group Areas Act and the Population Registration Act are in the process of being dismantled, the education system continues to be based on apartheid. There are still 19 education departments: five administering "white" education, one each for "Indians" and "coloureds", 11 for African education, and one umbrella department which controls the finances (money for education in the so-called independent homelands are channelled through the Department of Foreign Affairs), and sets the different "norms and standards" for the other 18, see Figure 1.

South Africa is also a society with deeply ingrained gender consciousness across all colour groups. Consequently, students are seldom exposed to anything other than gender biased education.

The civil engineering students at the University of Natal, in other words are the products of an education system best described as schizoid. In a class of 50, for example, it is conceivable that students have come from 18 different gender-biased, and "racially" separated education systems. This in itself places the students and the teacher at a great disadvantage. Some students are further disadvantaged either because of their colour or their gender. To give one statistic, in 1989 (the last year for which accurate figures

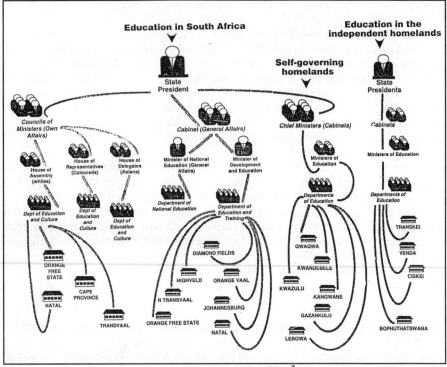

Figure 1: The structure of education in South Africa[2]

are available) per capita spending grew by R360 per white pupil compared with R191 for Africans.[3] In real terms this means that African students continue to be neglected particularly in the provision of basic science equipment, let alone computers. In so far as gender discrimination is concerned it cannot be quantified as statistics are not available on per capita expenditure on female students.

It should be pointed out that even students from privileged backgrounds have suffered the consequences of an incoherent and inchoate system. This is evident in the gaps that exist in their understanding of basic mathematics. In other words, under the South African education system all engineering students have been disadvantaged, though some more so than others. In reviewing courses, it became clear that from a structural engineering perspective, students suffer two main disadvantages: lack of spatial perception and lack of computer familiarity. This is especially so in the case of students from deprived educational backgrounds or where engineering is perceived to be a male domain.

The disadvantages are summarised in Table 1.

TABLE 1: DISADVANTAGES

Who	All students	Female	African students
Why	Apartheid education. Poor schooling in basic mathematics.	Gender-based education. Engineering perceived as male domain.	Poor health, housing and education. Instruction in second language (Afrikaans or English). Disjointed and interrupted schooling. No facilities.
What		Lack of spatial perception. Lack of computer familiarity.	

REMEDIES

The question then is how are the deficiencies to be remedied. One way to overcome the deficiency in spatial perception is to take a physical rather than a mathematical approach to teaching structural behaviour. To this end, courses have been reorganised to address the problem at an early stage. Another way to tackle spatial deficiency and computer unfamiliarity is to take a "functional computer literacy" approach as distinct from computer literacy. A third way has been to set assignments with a view to developing spatial and functional computer literacy skills. A fourth is to base assessment on the students' understanding of structural behaviour and not on their mathematical competence alone.

The objectives of the structures courses are hence:

- To develop an understanding of the physical behaviour of structures.
- To cultivate an intuitive appreciation of the structural behaviour based on qualitative considerations.
- To comprehend the mathematical representations of this behaviour.
- To provide access to quality structural software on suitable hardware and develop functional computer literacy.
- To interpret and validate the results and maintain a critical approach to the subject.

The NAFEMS guidelines are adhered to at all stages in the course development.[4]

SPATIAL PERCEPTION

There are three components to the spatial perception approach. The first is the integration of a spatial theme throughout the structural analysis courses as a prerequisite to the finite element course. A correct balance between physical behaviour, theoretical representation and numerical modelling is recognised as being essential to any finite elements course. By structural behaviour is meant:

4

- how loads are carried from their point of application to the supports (boundaries) where they are assumed to be dissipated;
- how the structure deforms under loading; and
- how the structure resists this deformation.

Indeed, the finite element course can be considered to start at least as early as the basic structures courses.

The second component of the spatial perception approach is the combining of physical and mathematical insights. Students lose a commonsense view of reality if the teaching of numerical analysis becomes remote and unrelated to the physical behaviour. The spatial perception approach is based on the premise that deformations and deflections can be seen and measured, whereas stresses and actions cannot. Consequently, students start with the observation and measurement of deformation, and derive the action and stress distributions. It results in a natural understanding of structural behaviour and gives the student confidence in tackling more complex finite element problems. This is in contrast to the past where physical concepts tended to be grasped less rapidly than the structural mathematics.

The third component of this approach is the development of an appreciation of deflected shapes and structural appropriateness, which are the qualitative indicators of structural behaviour. However, qualitative analysis is unlikely to be learnt as a by-product of quantitative analysis. Students better learn to understand the basic behaviour of structures when they are required to sketch the approximate deformation and stress distribution before analysis.

TABLE 2: STRUCTURE OF SPATIAL PERCEPTION APPROACH

Module 1	Strain at a point including measurement with strain gauges on the surface. Hence stresses can be derived from the strains.
Module 2	The strain over a cross-section surface is obtained by assuming the deformation pattern - typically that plane sections remain plane after deformation. Hence integration gives the distribution of stress across the section in terms of area, first moment of area, and second moment of area.
Module 3	The deformation along a one dimensional member is obtained by assuming the mode of deformation at each section along the member. Hence the actions at any point on the member can be derived.
Module 4	Introduction to structural theorems and principles: Virtual Work, Minimum Potential Energy, Reciprocal theorem, etc.
Module 5	Displacement of joints are found by assuming their connectivity and topology, and integration around the joint to give the member end actions. This allows for the displacement of frames to be determined, and hence member actions can be derived.
Module 6	Displacement of nodes in a continuum are found by assuming the displacement patterns between nodes. The finite element analysis determines the deformation of the material and hence the stresses at any point can be derived.

The dominant philosophy of the spatial perception approach is for students to gain practical experience in determining structural behaviour. It is considered more important for them to be familiar with the factors that influence the accuracy of the predictions and the means of assessing that accuracy, rather than to have a detailed understanding of theoretical considerations. Subsequently, the modules (Table 2) include a significant amount of practical work as well as qualitative assessment of the behaviour of structures based on assumed deformation patterns.

Case studies with known behaviour which are not trivial in their implications are employed, as for example, the deflection of a cantilever under load. Computer analyses are used so that students gain familiarity with computers. The best way to get deflected shapes, for instance, is through computer structural analysis. It should be stressed, however, that there is little point in using case studies that use dull repetitive calculations that can be done by a computer or ones that do not illustrate any new principle.

It is found that spatial perception and structural behaviour is better understood by students, both disadvantaged and advantaged alike, when they are given hands on practice with a quality structural analysis package. In this regard, a finite element package limited to about 50 elements is found to be sufficient. The qualitative analysis of structures is an important finite element prerequisite. Neglect of this aspect will produce technicians whose understanding of structural behaviour is poor when assessed by standards which are generally accepted by the profession. However, for a spatial perception deficiency to be remedied, merely training students to use a package is inadequate. It is also necessary to teach them functional computer literacy.

FUNCTIONAL COMPUTER LITERACY

A quality finite element package is an invaluable vehicle for remedying spatial inadequacy by aiding the development of an instinctive feeling for structural behaviour. Little, if anything, though is to be gained by continuing to teach the analytical techniques that were in fashion before computers came to be widely used by engineers.

Traditional finite element courses and texts are based on worked examples, most of which can be solved manually. They also focus on the development of FORTRAN programs to enable the student to obtain experience of finite element computations. Neither of these approaches give the student much idea of structural behaviour.

Students should not be required to take a course in computer programming and then find that they do not have to program until required to do so as part of a finite element course. Such programs are invariably simple and of unproven value. Students from disadvantaged backgrounds in particular have found the exigencies of computer programming a discouraging aspect of engineering courses. Moreover, the rate of development of computer hardware and software limits the usefulness of programming by students. As it is, in many areas of finite element education it takes considerable effort to keep the course content up to date by adding changing computer technology to the discipline.

Software appreciation and usage is, therefore, of greater value. It is preferable to use quality structural analysis packages to support the computational side of courses, rather than to have students develop their own programs. However, the use of a quality package should not be in isolation from the relevant theory, but be used to develop a critical attitude to mesh design and assessment of the accuracy of the results.

Since few specialists need to write a new element, the necessity to code a familiar element no longer exists. The costs of development, especially of quality code, are high compared with their use. For example:

$$\frac{cost\ to\ program\ one\ line\ of\ code}{cost\ to\ execute\ one\ line\ of\ code} = \frac{100\ million}{1}$$

It follows that it is more sensible to teach students the use of finite elements, rather than teach finite element development.

The term computer literacy usually means the reading and writing of computer languages and programs. The inference is that a computer literate person is one who can program. Yet, the unnecessary demands of programming often mean that sight is lost of the problem and its solution. It is to the detriment of the engineering profession to educate supposedly computer literate people who have had limited exposure to computing, and produce programs of unproven quality.

Functional computer literacy by comparison, teaches a student to read from and write "into" a computer. It does not require a knowledge of computer languages or the ability to write programs. It is functional because it concentrates on providing solutions to problems, structural or otherwise.

Functional computer literacy in finite elements requires a thorough grounding in finite element theory, an understanding of numerical methods, and familiarity with commercial packages and their uses. Part of the teaching of functional computer literacy is to ensure that students that come from disadvantaged backgrounds, and have not had access to computers even for word processing, are not discouraged from using finite elements because they are computer based.

Students should be exposed to the practical use of good quality code. The aim is not to go into algebraic detail of specific elements, or to train students in the use of a specific package, but to develop an understanding of structural behaviour and functional computer literacy.

One way to develop functional computer literacy is to start with the emphasis on pictorial solutions. In so doing, the student is educated to look at the output to judge whether it looks right or wrong, and know what must be done to correct the errors. When a student reaches a point where critical examination of graphical computer output is used to develop design, then it can be said that he or she has achieved an understanding of structural behaviour as it relates to functional computer literacy.

The availability of quality commercial finite element packages gives students the option of obtaining very detailed information about structural behaviour. However, it should not be assumed that it lessens the need for sound engineering judgement on the part of the student. On the contrary, an increased understanding is called for. For example, manual methods of structural analysis only consider bending deformations. But with computer methods, there's axial, shearing and torsional deformations. So the student needs to understand these deformation patterns as well.

There is little doubt, that computer based structural analysis courses are required to cope with the demands of the modern design office. It should be borne in mind that typical small jobs require only 20-50 elements, whereas larger jobs may have more than 1000 elements. In the work area the engineer is likely to encounter situations where there are jobs that require that the limitations of the package be exceeded. It is, therefore, better experience for students to learn with structural packages limited to 50 elements, for example, so that they learn at an early stage the ways by which they can circumvent the limits.

ASSIGNMENTS

Aims

Practical assignments are integral to the effective teaching of structural behaviour. These take the form of qualitative representation of the behaviour as a preliminary step to experimental measurement and computational analysis.

Assignments are designed to:

*   introduce the student to structural behaviour with a reliance on engineering, mathematical and intuitive skills;
*   develop functional computer literacy in structural behaviour;
*   show important concepts and procedures with simple examples; and
*   relate the techniques to reality.

The aim of these assignments is to guide users of finite element systems to achieve high quality results from analyses. The scope is limited to the linear analyses of structures and directly related problems such as thermal conduction and fluid flow.

Students learning manual methods of analysis build up confidence and understanding by working through simple examples that give a good approximation to the structural behaviour. This "back-of-an-envelope" approach is a prerequisite for effective use of the finite element method.

Analytical finite element examples do little more than reproduce standard solutions, and usually require tedious and mechanical matrix manipulations. This stems from the fact that the practical application of the finite element technique requires the use of computer programs. The successful applications of the finite element method to a variety of problems starts with education on the method, and practice in modelling it.

The assignments aim to teach students that the computer is simply a tool to solve problems and not an end in itself. They provide a breadth of understanding of structural behaviour that is not apparent from a single analysis, and the opportunity to evaluate and establish confidence in each method. The demand is made on each student to experience and test problem solving abilities on a continuous basis. This is done in a computing environment which can cope with finite element packages.

Good modelling is an acquired skill resulting from much practice and evaluation. The problem of poor modelling is hence given top priority. Modelling techniques form a central component of the assignments (as well as the course) and the student learns to interpret results in relation to a qualitative assessment. The opportunity is given to practice the art of modelling and to make comparisons. Modelling requires that several iterations of a solution are undertaken until an acceptable outcome is achieved.

Quality assurance is introduced in the assignments by an appropriate integration of theoretical, experimental and numerical studies using simple structures or components. Design courses are run in parallel to the theory, and the assignments are devised to flow into the design area.

The assignments reinforce the underlying principles introduced by lectures:

- The importance of understanding the behaviour before any modelling is attempted.
- The difficulty of relating numerical and physical models.
- The need for independent checks on the accuracy of the results.
- Inaccurate results can be obtained.
- Mesh refinement may improve the accuracy of the results.
- Higher order elements are more accurate than simple elements.
- The shape of the element effects the accuracy of the results.
- The displacement, strain and stress results are only approximate.

Implementation

The best way to familiarise students with structural behaviour is to give them hands on practice with the types of problems and challenges they might face after graduation. They are provided with case studies drawn from actual situations, and told to approximate solutions manually before computer calculation.

The assignments are chosen so that they:

- Build finite element concepts slowly, maintaining simplicity and clarity.
- Provide straightforward illustrations and comprehensive examples to give students a clear understanding of all concepts.
- Use colour to help visualise the concepts.

9

The assignments are simple which means that computational times are short, and the results readily available. The inaccuracies in the technique and the ways of improving accuracy can rapidly be investigated with guidance from the demonstrator.

The simple models that are used for the assignments are chosen for four main reasons.

- They are relevant to engineering practice.
- They have theoretical and analytical solutions.
- They are easy to understand.
- They can be tested against experimental results.

Problem specifications are provided for each assignment, together with an outline of the computer operations (commands and mouse movements) to achieve a solution. The students work individually on each assignment, for up to two weeks, and a report is submitted on completion. These reports are marked within a week to ensure that the students receive the feedback to enable them to improve as the course proceeds.

The notes accompanying the assignments provide guidelines for good practice, and contain sufficient information to enable the problems to be modelled and analyzed on a particular finite element system. The notes are intended to complement, but not supplant, the user manuals that form a part of all finite element systems.

Composition of assignments

There are nine computer assignments and two practical assignments, with two weeks allowed for the completion of each. Each assignment includes a set of guidelines which form part of a general checklist that can be followed in the solution of practical problems.

All assignments are open-ended. There are no correct answers to the finite element problems but, where possible, target solutions are given. These may be theoretical, experimental, or obtained from other finite element systems.

Particular assignments allow students to determine the capabilities of finite elements by:

- Making an initial guess at behaviour.
- Encouraging investigation of alternate meshes.
- Accuracy and convergence studies using progressively finer subdivisions.
- Comparison of finite element calculations with exact or known solutions for similar cases, such as benchmark tests.
- Checking methods for critical assessment of results.
- Using experience of previous calculations as a guide.
- Practical work testing a deep beam and a bridge deck.

The assignments are shown in Table 3.

TABLE 3: ASSIGNMENTS

Computer assignment 1	Problem definition - qualitative assessment of behaviour
Computer assignment 2	Basic geometry
Computer assignment 3	Element selection and idealization
Computer assignment 4	External actions
Computer assignment 5	Constraints and boundary conditions
Computer assignment 6	Adjacent structures and symmetry
Computer assignment 7	Solution methods
Computer assignment 8	Results selection, presentation and interpretation
Computer assignment 9	Planning, control and documentation
Laboratory assignment 1	Deep beam test
Laboratory assignment 2	Bridge deck test

Response of students

The structures course is compulsory. When students were told that it would be a computer orientated course, they were unhappy with the computing aspects. At the start of the course about 70 per cent indicated that they were fearful of using computers. It was apparent that this fear came from their initial exposure to computers during an earlier compulsory computer programming course.

Once introduced to the computing technique through simple assignments using interactive, user friendly software on the PC, students tackled more advanced assignments with greater confidence. They learnt computer usage, finite elements and structural behaviour as the course progressed; even those who had very limited computer experience rapidly demonstrated computer proficiency. They became enthusiastic about their new capabilities as their appreciation of the computer grew. Even teething problems with hardware/software were regarded as an important part of their learning experience, and as being more of a nuisance than a hindrance.

Students gained valuable experience in input and interpretation of the results. They appreciated their reports on assignments being returned to them within a week. The reports were an indispensable component as they provided an opportunity for the students to present their results and assessment of the different elements and meshes. The ready feed back enabled the detection and correction of bad habits and wrong perceptions. Considerable improvement in finite element functional computer literacy was evident.

The students gained practical experience of:

- Problem description through keyboard and mouse operations.

11

- Procedures for generating meshes, specifying boundary conditions, constraints, material properties and applied loads.
- Selection, presentation and interpretation of results.
- Computer hardware and software, particularly graphics.

The response of students to the finite elements course is favourable. They felt it gives real hands on experience, and considerably improves their structural understanding. They found the course material stimulating, if demanding, and were particularly impressed by the comprehensive "Help" facilities, which precluded the need to refer to the user manual.

One particular response that needs further comment concerns the learning pattern of two different student groups. One was a group of graduate engineers with several years design and computer experience. They made use of the finite element package in their own design offices, and therefore worked separated from each other. The second group was made up of undergraduates who worked individually on computers located in the same room, but nonetheless discussed the work among themselves. Although the two courses were run in parallel, the undergraduates learnt considerably quicker than the graduates.

This response indicates that students learning finite elements do so better when working individually within one location such as a computer room. That is not to say, however, that they should work in teams as such an approach further disadvantages those who are unfamiliar with computers, and thus rely on those with more experience.

ASSESSMENT

The traditional approach to structural analysis concentrates on the development of mathematically based analytical skills. Students are expected to satisfy a form of assessment based on numerical tests, without necessarily understanding the behaviour of the structures being analyzed or the computer based systems on which finite elements are implemented.

The question of assessing a student's understanding of finite elements is not straightforward, particularly with regard to examinations. It is a decidedly more difficult task to inculcate in a student an appreciation of structural behaviour, than it is to get a student to answer examination questions on any given problem or technique. Examinations alone are not suitable for assessing a student's knowledge of finite elements.

A more balanced approach to assessment is, therefore, needed whereby the student's functional computer literacy in finite elements is monitored during the progress of the course, and the understanding of structural aspects assessed by examination. On the one hand, assignments need to be evaluated to give suitable feedback to the student, and provide part of the final course mark. On the other hand, the examination needs to be more qualitative than quantitative to provide a good balance between theory and practice.

The compulsory assignments are central to the learning process. It is doubtful, however, if assessment based solely on the operation of a finite element system is sufficient to indicate the level of the student's understanding of structural behaviour. It is also necessary that the student be able to provide a plausible physical explanation of the behaviour. The requirement to do so under examination conditions gives confidence that the student is competent to take responsibility for the application of finite elements for the analysis of real world problems.

The assessment of the finite element course is hence in two halves: continuous assessment and examination. The continuous assessment is based on the assignments which are compulsory. Students are assessed on their functional computer literacy, presentation of results and understanding of structural behaviour. Extra marks are awarded for ingenuity and investigation beyond the scope of the assignments. Marks are deducted for lateness and lack of originality. Thus the mark achieved does not necessarily reflect the effort devoted to the assignment, but rather the understanding of all aspects of the problem. The continuous assessment is worth 50 per cent of the final mark.

The examination is a three hour open book one, in two sections. The first section, which is worth 30 per cent of the examination, requires the development of suitable guidelines for some aspect of finite elements, such as for the creation of a finite element mesh. The second section, which is worth 70 per cent, requires the development of a suitable finite element model of part of a structure such as a twin box beam bridge deck. The examination paper and the scripts are checked by an external examiner.

The marks achieved by students did not reveal any significant disparity between the different assessment approaches. Neither was there any discernible difference between the marks attained by those from a disadvantaged background and those who were not.

CONCLUSIONS

The teaching of finite elements shows that disadvantages such as lack of spatial perception and lack of computer familiarity can be overcome through hands on practice with a quality finite element package, assignments that are continuously assessed, and an examination based on the understanding of structural behaviour. It is not, however, a panacea for the ills of a discriminatory education system. The main responsibility for redressing past educational injustices lies with the government. For the present, however, it does not seem likely that resources will be made available to further assist deprived engineering students. The responsibility, therefore, falls on tertiary institutions and the engineering profession to ensure that provision is made for those from a disadvantaged background.

Many tertiary institutions have instituted what is known as a "bridging year" to help students make the transition from school to university, and overcome the disparity between the different education systems. In engineering, students are brought up to first year university standard in the core subjects. The funds for the bridging programme come mainly from the private sector. Disadvantaged students further receive assistance through student support schemes run by tertiary institutions where individual tuition is given in subjects where students have difficulties, such as mathematics and physics.

There is room, however, for further improvement. Students from financially deprived backgrounds, for example, are often asked to come long distances, and at great cost for interviews that are often unnecessary. On a another level, there is a need actively to encourage women and Africans to engage in post-graduate work to provide different insights to the discipline.

The engineering profession too has a role in the development of graduates from a disadvantaged background, particularly by encouraging involvement in professional society activities. There is a notable absence of women and African engineers within professional circles, and they are inadequately represented in decision making positions. It is in the long term interests of the profession to recruit and retain engineers from disadvantaged backgrounds and provide the support for their professional development.

The neglect of human resources through a discriminatory education has resulted in a severely imbalanced society. Engineering, in as much as any other profession, has the responsibility to ensure that there is a just and equitable distribution of skills that are critical to the building of a New South Africa. To do so requires the will to remedy past injustices and the vision to provide opportunities for the future.

REFERENCES

1. Weinberg, S., State does the blackboard crab-walk. In The Weekly Mail, 15-21 February 1991, p. 41.

2. In The Weekly Mail, 15-21 February 1991, p. 42.

3. South African Institute of Race Relations, Race Relations Survey 1989/1990, Johannesburg, pp 147-52.

4. Guidelines to finite element practice, NAFEMS, Glasgow, 1984.

REQUIRED KNOWLEDGE FOR FINITE ELEMENT SOFTWARE USE

IAIN A MACLEOD
Department of Civil Engineering
University of Strathclyde

ABSTRACT

The paper discusses the knowledge spectrum for the finite element method and concludes that few of the details of processing are required by users of FE software.

INTRODUCTION

As Information Technology develops, a problem that we face is to decide to what extent details of the computing processes need to be visible to the user. In electronics, for example, integrated circuits have greatly reduced the need for circuit designers to deal with components. We can now treat a finite element package in the same way as an integrated circuit. We need to know only the general principles of how input is transformed into output. Within the information explosion there seems little point in gaining knowledge that will not be used.

This paper discusses the limits of useful knowledge for users of finite element software.

The purpose of finite element analysis is to predict the behaviour of physical systems. The physical system to be modelled is described here as the "prototype" and the person responsible for creating the model is the "user".

The discussion relates to the use of structural finite elements but the general principles apply in other applications.

STEPS IN THE FINITE ELEMENT METHOD

These are:

1. Define the constitutive relationships
2. Define the element relationships
3. Form the system model

4. Solve
5. Back substitute

These steps are considered separately in terms of assumptions and of processing.

CONSTITUTIVE RELATIONSHIPS

These represent the primary definition of material behaviour. Knowledge of the basic assumptions used in deriving them and how they relate to the behaviour of the prototype are essential user requirements.

Consider, for example, the constitutive relationship for uniaxial bending:

$$EI\frac{d^2v}{dx^2} = M \qquad (1)$$

This is commonly used in engineering but I expect that many people who do use it regularly could not give a coherent statement about the basic assumptions (i.e. plane sections remain plane, small deformations, no shear deformation) and their limitations. Users should be able to give such statements about any constitutive relationship they use.

Processing
It is worthwhile to learn the details of derivations of simple relationships such as bending and plane stress but such knowledge tends not to be retained for long. Detailed derivation of the constitutive relationships for non-linear analysis are more complex and would not normally be required for users. However, the basic assumptions and their validities should be clearly understood.

User manuals for software should give reference to where such information can be found for all constitutive relationships used.

ELEMENT RELATIONSHIPS

I divide elements into two classes (1):

Type A are those which require no further assumptions beyond the constitutive relationship to establish the element stiffness matrix. The uniform bar (axial force), bending and torsion elements (which together form a uniform beam element) are the only members of this class but in structural engineering they are the most commonly used. It is most important to realise that these elements do not need mesh refinement to improve accuracy. Strictly speaking, assumptions are made about the deformed shape of these elements but the functions used satisfy the constitutive relationships and therefore no approximations are involved.

For **Type B** elements, further approximations are required and mesh refinement is needed to improve accuracy. An important question is: "What is the desirable extent of user knowledge for these assumptions?" The assumptions for basic plane stress elements (constant strain or linear strain) are easy to understand and can be helpful for appreciating behaviour but knowledge of the assumptions for more complex elements (e.g. hybrid elements) does not help the user to understand how the element will behave. In fact information about such assumptions can be difficult to obtain. The main plate bending element in the LUSAS system (2) gives good results but the details have not been published to my knowledge.

In reference (3) Irons and Ahmad make a very revealing statement. In relation to the Semi-Loof shell element they say "We are convinced that few workers will be motivated to enquire in any depth about the Semi-Loof shell element. Indeed, we ourselves find the original reference logical but extremely taxing to read." If Bruce Irons found his own paper difficult, how many users would be able to understand the derivation even if they tried? I did meet someone who claimed to have got down to the basics of the Semi-Loof element but such expertise is rare.

Knowledge of the performance of Type B elements comes mainly from observing their behaviour in use. Users should develop "favourite elements" and get to know them well.

Convergence studies are essential for this purpose. The NAFEMS (4) Benchmark tests are designed to test element performance. They can also help users to develop a feel for element behaviour. Again, user manuals should quote, or give reference to, convergence studies for all Type B elements.

When looking at convergence, it is important to review the range of output parameters. For example, some elements can give very good results for deformation and direct stress but give poor predictions for shear stress.

Processing

Having taught students for over 20 years about how element matrices are set up using $k = \int B^T D B$, the use of local co-ordinates, and change of variable and numerical integration, I have come to the conclusion that this is not necessary user knowledge. It is of course essential for those who want to write element routines but what proportion of students who are taught about finite elements are ever in this situation? Fewer than 1% nowadays, I expect. Those who do need it can develop the necessary knowledge as required. To demonstrate this point, consider the concept of the Jacobian matrix J used in the transformation of variables. I can use this concept in a black box mode but I have no 'feel' for why one needs to multiply an integral by the determinant of J when the variables are changed. The concept therefore does nothing for my understanding of element behaviour and so I can happily do without it as a user.

THE SYSTEM MODEL

The assumptions for the system model beyond that for the elements include:

1. Definition of the boundary conditions
2. Definition of loading.

The user must have knowledge of how the behaviour of the system model relates to that of the prototype, where to refine the mesh, and the validity of the imposed boundary conditions. The validity of the loading specified for the model in comparison with that which the prototype will experience is of course a major issue but can be considered to be separate from the modelling of the prototype itself.

The concept of conforming and non-conforming elements is important. That an assemblage of conforming elements provides a lower bound to the exact solution (i.e. the solution one would get if the constitutive relationship, equilibrium and compatibility were all satisfied by the model) for an influence coefficient is important. Such knowledge is useful in checking since the sign of error can be an indication of validity. The reasons why elements do or do not 'conform' is however of lesser value to the user. User manuals should provide information about whether or not elements are conforming.

Processing
The process of forming a structural stiffness matrix need only be understood in principle. I was an early user of the direct stiffness method but this is not helpful in understanding the principles involved. It is probably easier to understand the principles from the connection matrix approach (1,5) where the structural stiffness $\mathbf{K}$ is defined by

$$\mathbf{K} = \mathbf{C}^T \mathbf{k}_e \mathbf{C} \qquad (2)$$

where $\mathbf{C}$ is the compatibility (connection) matrix (consisting of 1's or 0's) which equates element deformations with system deformations,
$\mathbf{C}^T$ is the corresponding equilibrium matrix, and
$\mathbf{k}_e$ is the diagonally partitioned matrix of element stiffnesses.

While one would not use such a transformation in a program, the development of equation (2) does give some insight into the principles involved in setting up $\mathbf{K}$.

SOLUTION OF SYSTEM EQUATIONS

This is the major processing component of the Finite Element Method and again the user needs to understand general principles rather than have detailed knowledge of techniques.

Gaussian Elimination, which is normally used for linear solutions, is taught in schools so most users do have a basic appreciation of this process. The basic principle of decomposing to a triangular form and using this to provide the solution is common to all direct solution methods and users only need an appreciation of one of them, i.e. Gaussian Elimination.

Direct solutions use band or wavefront methods. The former needs the node numbering to be optimised; for the latter, the element numbering affects the efficiency. Most FE packages reorder nodes or element numbers as appropriate but where this is a user requirement only the rules for such ordering need be known.

For non-linear solutions some user control over the load or time increments etc can be required. Here again details of the process need not be visible. Guidelines plus experience in use of the system are the main requirements for developing knowledge about increment choice.

An important feature of the solution process is the potential for ill-conditioning. Users need to understand:

o   the numerical source of ill-conditioning

o   typical situations in which this can occur

o   how to detect its presence

BACK-SUBSTITUTION

Since stresses are often quoted at Gauss points users need to know where they are located. Beyond that 'needed in the numerical integration' is the limit of required knowledge about Gauss points.

PROGRAMMING USING SUB-ROUTINES

Some teachers have their students write FE software using subroutines (such as those from the NAG Finite Element Library software (6)). If the objective is to develop better understanding of the processing steps then such activity is advantageous. I submit that such understanding is not a user requirement.

POST PROCESSING

A main question by the user is "How well do my results represent behaviour of the prototype?" To answer this, the user needs to understand the behaviour of the prototype. But the purpose of the analysis is to gain knowledge of this behaviour; therefore we appear to have a 'Catch 22' situation.

The way out of what appears to be an inward turning problem is to view the purpose of the analysis not only as a means of solving the particular problem in hand but as a potential source for more general understanding of behaviour of the prototype.

Two techniques which help to develop this understanding are now discussed.

Use of a checking critique

Detailed appraisal of the results of a run can often reveal important trends in behaviour. One has to make assumptions about behaviour and relate them to the results. If there is a match, one's interpretation of behaviour may be correct (although one must accept that correlations may be close for the wrong reasons). If there is no match then either

•   there are errors in the analytical model, or
•   one's understanding of behaviour of the model is wrong.

If the former situation pertains then the errors can be identified and corrected. In the latter situation, if the behaviour can be explained, then potentially valuable knowledge about behaviour will be gained.

Parameter studies

One can develop knowledge of behaviour by altering the model and observing the effects of the changes. This is a potentially powerful technique.

It is my experience however that many students are happier with procedures than with investigation of behaviour. Whether this is a function of their experience in education which has concentrated on procedures or due to a genetic inclination is not possible to know. I suspect however that our educational system does breed procedure-dominant rather than understanding-dominant attitudes.

CONCLUSION

A recent paper of the Engineering Professors Conference (7) suggests that a new balance should be struck in engineering education development. Understanding should be given higher priority. This needs time and some knowledge and skill will need to be sacrificed. We are captives of an experience and, having been brought up in an environment where the details of processing were important, it is difficult to accept that knowledge of the process may now have a low priority or may even be irrelevant for finite element users.

My contention is that users can cast processing aside to concentrate on the behaviour of the analytical model and its relationship with the prototype.

REFERENCES

1. MacLeod I A. Analytical Modelling of Structural Systems. Ellis Horwood, 1990.

2. LUSAS Finite Element Stress Analysis System. Finite Element Analysis Ltd, London, 1987.

3. Irons B & Ahmad S. Techniques of Finite Elements. Ellis Horwood, 1980.

4. NAFEMS, National Engineering Laboratory, East Kilbride, Glasgow.

5. Hall A S & Woodhead R W. Frame Analysis. Wiley, 1961.

6. The National Algorithms Group (NAG), Finite Element Library, 1987.

7. Engineering Professors Conference. The Future Pattern of 1st Degree Courses in Engineering, Occasional Paper No 3, Feb 1991.

INDEX OF CONTRIBUTORS